A Primer of Conservation Behavior

A PRIMER OF CONSERVATION BEHAVIOR

Daniel T. Blumstein and Esteban Fernández-Juricic

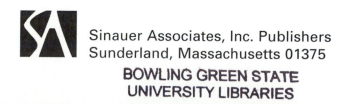

Sinauer Associates, Inc. Publishers
Sunderland, Massachusetts 01375

About the Cover
Populations of endangered Seychelles warblers
(*Acrocephalus sechellensis*), black-footed ferrets (*Mustela
nigripes*), kakapo (*Strigops habroptila*), giant pandas
(*Ailuropoda melanoleuca*), and California condors (*Gymnogyps
californianus*) have all benefited from insightful applications
of behavioral principles into their management. These and
other conservation behavior successes are discussed in this
book. Illustration: pen and ink drawing by Gabriela Sincich
(http://gsincich.com).

For information, address:
Sinauer Associates, Inc., 23 Plumtree Road, Sunderland, MA 01375 USA
Fax: 413-549-1118
E-mail: orders@sinauer.com; publish@sinauer.com
Internet: www.sinauer.com

Library of Congress Cataloging-in-Publication Data

Blumstein, Daniel T.
  A primer of conservation behavior / Daniel T. Blumstein & Esteban Fernández-Juricic.
      p. cm.
  Includes bibliographical references and index.
  ISBN 978-0-87893-401-0
  1. Animal behavior. 2. Wildlife conservation. I. Fernández-Juricic, Esteban, 1971- II.
Title.
  QL751.B56 2010
  639.9—dc22

                              2010025756

Printed in U.S.A.

5 4 3 2 1

To David and Melissa.
May you have the opportunities to experience
and enjoy the wonders of nature that we have had.

Table of Contents

Preface

Conservation behavior is the application of knowledge of animal behavior in order to solve wildlife conservation problems. We wrote this *Primer* because we felt that there was a need to nurture the development of biologists interested in using specific animal behavior conceptual and methodological tools for solving biological conservation and wildlife management problems. While there are several excellent reviews and edited volumes that discuss the integration of behavior and conservation biology, there has been no practical guide fostering integration and showing how to apply some methodologies to issues that would benefit from an animal behavior perspective. This book is broadly aimed at biologists interested in or practicing conservation biology and wildlife management (undergraduate and graduate students, conservation biologists, ecologists, wildlife managers, zoo personnel, animal behaviorists, and behavioral ecologists). This *Primer* will help familiarize the reader with some conceptual frameworks of the behavioral literature that are relevant for helping to save endangered or threatened species or to manage wildlife, and it will help the reader understand the constraints of conservation in practice. If you are a student, the *Primer* can be read in conjunction with textbooks on animal behavior (e.g., Alcock 2009) and conservation biology (e.g., Primack 2008; Groom et al. 2006). It can also serve as a core reading for an advanced undergraduate or graduate seminar in conservation behavior. In addition to citing papers, we cite useful computer programs, Websites, and other resources that can help you apply specific methods to solve specific (but not all) conservation problems. In the end, we believe that providing some guidance on specific theoretical and methodological approaches to bringing conservation and behavior together will encourage the development of even better methods to solve existing and future conservation and wildlife management problems.

Conservation behavior is a burgeoning field. We have selected a *subset* of topics for this *Primer* that draw links between how animals respond to anthropogenic impacts and how we can use a mechanistic understanding of behavior to facilitate coexistence between wildlife and humans. Much of conservation behavior focuses on single-species management. Thus, we need to know about a species' social behavior, reproductive behavior, and antipredator behavior. When a species is extremely rare, we may need to breed it in captivity (this is often a real challenge) and then reintroduce it to the wild (also a challenge). Thus, a detailed understanding of behavioral processes may help increase the success of translocations and reintroductions. Additionally, there are many overabundant species that may require some control, and conservation behavior can provide a framework to help improve some current techniques or a toolkit to develop novel techniques. Whether the target species is rare or abundant, we must know what behavioral factors influence its population size (e.g., predation, survival, reproductive skew) to be able to better predict density changes over time under different management scenarios.

This *Primer* only touches the surface of the behavioral topics that can be useful for conservation and wildlife management. We hope that those stimulated by the topics discussed here will be encouraged to learn more about behavior, conservation, and wildlife management and to help develop additional links that can be explored in the future. We have set up a Web site (www.eeb.ucla.edu/APrimerOfConservationBehavior/index.html) for comments and opinions, and we welcome your suggestions.

The field of conservation behavior is relatively young, and it has largely been defined by an enthusiastic cohort of young behavioral ecologists and conservation biologists. We thank Lisa Angeloni, Marc Bekoff, Rich Buchholz, Tim Caro, Colleen Cassidy-St. Claire, Kevin Crooks, John Eadie, Guillermo Paz-y-Miño, Bruce Schulte, Deb Shier, Judy Stamps, Ted Stankowich, Ron Swaisgood, and Pat Zollner for ongoing conversations about conservation behavior and for their encouragement to write this *Primer*. We have learned a lot from conversations with Peter Arcese, Joel Berger, Guy Cowlishaw, Scott Creel, Adrian Delnevo, Jenny Gill, Jan Komdeur, Georgina Mace, Ian McLean, Gary Roemer, Martin Schlaepfer, Tom Smith, Bill Sutherland, and Bob Wayne.

Acknowledgments

We thank Ted Stankowich for comments and discussions on previous incarnations of parts of the book; Lisa Angeloni, Brad Blackwell, Lauren Brierley, Tim Caro, Kerry Rabenold, Dan Stahler, and Bill Sutherland read the entire manuscript, and we are extremely grateful for their comments that helped us improve the book. We thank Gabriela Sincich for her original cover art and design, Neil Losin for letting us use his picture of a sage grouse lek, Steven Frischling and Sarah Partan for the photograph of the robotic squirrel, and Gail Patricelli for her photograph of the robotic sage grouse.

We thank Andy Sinauer for sharing our belief in the need for a *Primer*; Lou Doucette for copyediting; Production Editors Sydney Carroll and Julie HawkOwl for helping this *Primer* see the light of day; Joanne Delphia for rendering the art; and Christopher Small and Janice Holabird for their work on the cover and design of the book.

Finally, we thank Janice and David, and Gabriela and Melissa for tolerating our absences while we were researching and writing this *Primer*.

Daniel T. Blumstein
Los Angeles, California, and Gothic, Colorado

Esteban Fernández-Juricic
West Lafayette, Indiana
February 2010

About the Authors

Daniel T. Blumstein is a Professor and Chair of the Department of Ecology and Evolutionary Biology at the University of California, Los Angeles. He received his undergraduate degrees in Environmental, Population, and Organismic Biology and in Environmental Conservation at the University of Colorado, Boulder. He received his PhD in Animal Behavior at the University of California, Davis, and was a postdoctoral fellow at the University of Marburg (Germany), the University of Kansas, and Macquarie University (Australia). He has studied behavior and conservation in Australia, Canada, the Caribbean, Germany, Kenya, New Zealand, Pakistan, Russia, and the United States. He has served on endangered species recovery teams and is a member of the IUCN Reintroduction Specialist Group and the Conservation Behavior Committee of the Animal Behavior Society. He is a past editor of the journal *Animal Behaviour*, and is presently an associate editor of *The Quarterly Review of Biology*. He is also on the editorial boards of *Behavioral Ecology* and *Biology Letters*. He spends his summers studying marmot behavior and ecology at the Rocky Mountain Biological Laboratory, Gothic, Colorado.

Esteban Fernández-Juricic is an Associate Professor of Biological Sciences at Purdue University. He got his undergraduate degree at Universidad Nacional de Córdoba, Argentina. He received his PhD in animal ecology at Universidad Complutense de Madrid, Spain. He was a postdoctoral fellow at the University of Oxford (United Kingdom) and the University of Minnesota. Before his current position, he was an Assistant Professor at California State University, Long Beach. He has studied behavior and conservation in Argentina, Spain, the United Kingdom, and the United States. He is an associate editor of the *Journal of Applied Ecology* and *Behavioral Ecology and Sociobiology*. He is a member of the Conservation Behavior Committee of the Animal Behavior Society. He is currently interested in the integration of sensory ecology, behavioral ecology, and conservation biology.

1 What Is Conservation Behavior?

Conservation behavior, the application of knowledge of animal behavior to solve wildlife conservation problems, is a relatively young, integrative field. Biologists using conservation behavior apply theoretical and methodological insights from animal behavior, ethology, and behavioral ecology to conservation biology and wildlife management (**Box 1.1**). While conservation biology was created in the 1980s as a "crisis discipline" aimed at conserving biodiversity (Soulé & Wilcox 1980), it has matured into a genuinely integrative discipline (see Box 1.1). By contrast, conservation behavior formally emerged as a discipline with the publication in 1997 of Clemmons & Buchholz's edited proceedings of a workshop held at the 1995 Animal Behavior Society meeting in Lincoln, Nebraska. This edited volume was the first of four (Caro 1998; Gosling & Sutherland 2000; Festa-Bianchet & Apollonio 2003). It was joined by several substantive reviews (Strier 1997; Sutherland 1998a; Reed 1999; Anthony & Blumstein 2000), along with special issues of journals that were devoted to how the fields of behavior and conservation biology can be integrated (*Oikos* Volume 77, 1996; *Environmental Biology of Fishes* Volume 55, 1999; *Applied Animal Behavior Science* Volume 102: 3–4, 2007).

Despite this academic interest, recent critics have noted that behavior has little to offer to conservation biology, and that there have been only a few successful applications of behavioral knowledge to conserve or manage species (Caro 2007). Conservation behavior typically focuses on the conservation and management of single species. In a perfect world, we would preserve habitat and ecosystems and we would have no single-species management problems. However, the world is far from perfect and single-species management is an unfortunate fact of conservation. Many governments have legislation that focuses on single-species management (e.g., the United States Endangered Species Act; see Box 1.2), and sometimes the public demands it (e.g., consider public interest in pandas, whales, and elephants).

In this *Primer* we hope to show you specifically *how* behavioral knowledge is used (and can be used) to help conserve and manage threatened or vulnerable species, as well as control overabundant species, and how knowledge of the behavior of groups of single species may also be useful for multi-species management. The goal is to develop a toolkit that provides new approaches when facing conservation and management problems. We provide worked examples to illustrate precisely how biologists can apply behavior to solve management problems. Our audience includes both students of animal behavior and conservation biology as well as professional wildlife managers.

BOX 1.1 Some Important Definitions

Animal behavior is the formal study of non-human behavior. It seeks to develop general principles of behavior and predictive models.

Behavioral ecology is the economic study of animal behavior that focuses on behaviors' costs and benefits. It emerged in the late 1960s and initially focused on the study of the adaptive utility of behavior and developed "optimality models." Modern behavioral ecologists use a remarkable variety of tools and ask both proximate and ultimate questions (see Box 1.3) in their quest to understand the diversity of behavior.

Conservation biology emerged in the 1980s to conserve biodiversity (Soulé & Wilcox 1980). It has evolved into a mature and interdisciplinary discipline that combines "principles of ecology, biogeography, population genetics, economics, sociology, anthropology, philosophy, and other theoretically based disciplines to the maintenance of biodiversity" (Groom et al. 2006).

Ethology is the naturalistic study of animal behavior pioneered by European behavioral biologists in the middle part of the 20th century. Its initial focus was on causation and development, but modern ethologists ask both proximate questions—those about development and immediate causation—as well as ultimate questions—those about current adaptive utility and evolution (see Box 1.3).

Wildlife management is the application of scientific principles to conserve and manage wildlife populations. It has a long history of applying population biology tools and, more recently, to applying population genetic tools.

Why Is Conservation Behavior A Unique Field?

One might initially think that conservation biology explicitly integrates behavior into it and therefore it is not necessary to formally define conservation behavior. However, a formal recognition of animal behavior as being a valid component of conservation biology is important because it allows us to narrow down the questions conservation behavior can answer, which leads to the development of more specific methods and tools. Of course, maintaining animal biodiversity is only a subset of the biodiversity and ecosystem processes that conservation biology aims to conserve. With respect to managing single species (the realm of conservation behavior), population biology and population genetics are described as disciplines that have important insights. While ex-situ conservation (e.g., captive breeding), reintroduction (moving animals from captivity to the wild to restore or rescue a population), and translocation (moving animals from one wild location to another to restore or rescue a population) are discussed in conservation biology books, understanding the behavioral aspects of these tools can improve the effectiveness of conservation interventions.

Probably the closest daily practitioners of conservation behavior are wildlife biologists who deal with a multitude of very specific problems affecting single or multiple species. Wildlife biologists have formidable empirical knowledge of the behaviors of species they work with. Interestingly, wildlife managers are not typically trained in behavioral biology. To be a "certified wildlifer" (www.wildlifesociety.org), one must take 36 semester hours of Biological Science courses such as wildlife management, wildlife biology, ecology, zoology, and botany (as well as other requirements). Interestingly, a behavior course is only one of many possible options. By contrast, most people formally trained in behavior, are Ecology and Evolutionary Biology majors who take a lot of chemistry, physics, and biology, but are typically not required to take wildlife management. Although behavioral biologists may have received a more conceptual animal behavior training, they do not necessarily have a good understanding of field biology. Creating opportunities for conversations between wildlife biologists and behavioral biologists should lead to a fruitful integration of ideas. One example is population viability, which is to an extent mediated by behavior: fecundity by mating behavior, survival by foraging behavior and antipredator behavior, etc. Our aim in writing this *Primer* is to help foster these insights and conversations.

We should note that this simplistic view of training traditions (wildlife biologists and behavioral biologists) is not always that clear, and that there are many biologists that actively use behavior and many that do not. But, if you are interested in behavior or if you are passionate about the conservation of biodiversity, why not integrate conservation and behavior when appropriate? Since there are some challenges to integrating conservation and behavior, we will explore two answers to this question.

I am too busy doing cutting-edge behavioral research to think about conservation behavior

Many academics believe that by focusing on academic research they are unable to also think about conservation behavior or work with endangered species. To some extent, the same argument was used in the 1980s about why ecologists could not think about conservation biology. Fortunately, we now have many academic biology departments that embrace conservation biology, and many academic biologists are engaged in important conservation projects.

The arguments then, and now, often boil down to noting that since there is limited time to work, one must focus on what they need to do to get hired and promoted. It can also be argued, that to conduct cutting-edge research one needs to work with common "model" species and manipulate them to identify causality. Such options are sometimes not available with endangered species, which may be exceedingly rare, and may have additional constraints on experimentation. If sample sizes are small (because of rarity) it may be difficult to produce unambiguous answers. Moreover, some argue that when it comes time to search for a job in a non-applied discipline, or be promoted in a more theoretically-focused department, work with endangered species will be counted against them because they have taken the time to do this rather than publish another paper in a top-ranked theoretical journal.

However, conservation behavior embraces a variety of academically exciting topics. In this *Primer* we aim to show you how fundamental knowledge of behavioral mechanisms and evolution can provide useful information that can be used to maintain biodiversity. We hope to show you how theoretical discoveries in the fields of habitat selection, foraging, and antipredator behavior, communication and environmental acoustics, individuality and personalities, social behavior, and sexual selection have important implications and lessons for species management. And we hope to show you how some of these discoveries were made with no particular applied problem in mind but that does not prevent them from potentially being useful to solve conservation and management problems.

Thus, by doing cutting edge behavioral research, and particularly research that has demographic consequences, a broader impact of your work is that you are generating knowledge that could be used by a biologist working in conservation. To really help, focus on what you know best and think outside the immediate box. Then ask yourself, what are the demographic consequences of your behavior of interest? This is an essential step in developing individual based models (discussed in Chapter 2). Clearly write these implications in your academic publications. Persuade your editors to allow you to include these comments. By doing so, you are playing an important role of creating and disseminating knowledge that may be used in the future.

But, there are more things you can do. We suspect that you got interested in studying behavior because you are excited about animals and nature. Well, this may be your opportunity to contribute to saving the nature we value. To do so, you may take a few more steps.

For instance, have an honest discussion with those working in the front lines of protecting and managing wildlife about the problems they face. Learning about wildlife management needs is essential because there are problems that must be solved. By clearly understanding conservation and management needs, you might be able to build intellectual links between your area of behavioral expertise and a particular management solution. You may get excited about a particular national or international conservation topic or species of conservation concern and may wish to contact people in charge of that program and talk with them.

Actively seek collaborations with biologists working in wildlife management and conservation. These collaborations could take the form of research projects or even education projects. Bringing behavior into the picture can easily engage people as they generally enjoy learning about animals. Keep in mind that simply because you are working with a conservation question does not necessarily mean that money to fund the research suddenly appears. Often, it does not.

What about the question of where to publish conservation behavior research? Sometimes, conservation actions and management procedures create experimental situations that can be used to test important theoretical ideas. For instance, a conservation intervention on Seychelles warblers (*Acrocephalus sechellensis*) was used to test habitat saturation models and was ultimately published in the journal *Nature* (Komedeur 1992). Many other conservation behavior papers are published in top-ranked journals and we discuss a number of them in this *Primer*.

For all of the above reasons, we believe that theoretically inclined biologists who want to contribute to conservation behavior can do so. If we do not work to conserve the biodiversity we study, there will not be anything left to study.

I am too busy conserving to think about conservation behavior

In some circles, there is stigma about the role of animal behavior in conservation as having little or no value. Animal behavior is sometimes considered too academic or too elitist. The obvious implication is that some behavioral biologists are sometimes seen as disconnected from conserving and managing species in the real world. This may be true in some cases (see above). However, it is also true that often the reality of dealing with a species does call for knowledge of animal behavior.

Consider this example: During their long migration routes, migratory species travel across different landscapes, including urban areas. When crossing downtown areas with tall buildings, some birds collide with the windows and die. This source of mortality is estimated to be as high as 1 billion birds per year, and could reduce the local abundance of many species of conservation concern (Klem 1990). Despite the population level consequences, this can be seen as an animal behavior problem. Why can't birds avoid colliding with the buildings? Is it a problem with the ability of the bird to detect the building? Or perhaps the building is easily detected but the animals are attracted to it? Answering these fundamental questions requires using animal behavior concepts and tools (e.g., sensory biology, preferences, etc.). More importantly, answering these questions may bring a novel approach to reducing or eliminating this problem (e.g., making buildings more distinctive to birds, reducing night lighting to decrease attraction, etc.).

The key is thinking about the core of your conservation problem. Many problems may not point towards behavior at all. But, as you will see from the examples in this *Primer*, behavior is at the root of many problems. Identifying a fundamental animal behavior question that underlies your conservation problem may provide new insights on solving the problem because the field of animal behavior has a wide conceptual breath. This means that you can use animal behavior to put the observations you have already made in the field in an alternative, conceptual framework that can provide some guidance on how to manipulate experimental factors that will attract or repel a species, for instance. Using any kind of conceptual framework (be it behavior, or population ecology, or genetics) can lead you to the solutions more quickly than going through a trial-and-error process, which can sometimes be expensive, of what to modify. For instance, the framework of giving-up densities can be used to improve the suitability of patches for some species by reducing three types of costs that animals face: metabolic, missed foraging opportunities, and predation risk (see Chapter 6).

So, you may have seen the importance of behavior in some conservation problems and have identified a fundamental animal behavior question underlying your conservation problem. Does this mean you need to start reading books about animal behavior and behavioral ecology? You may do that, but you may also communicate with behavioral biologists. You would be surprised by how many academics started their careers studying behavior, but at a later stage in their professional career become deeply involved in conservation and management. These are the people who would be more willing to talk to you about your conservation problem. Google them, email them, phone them.

Your communication with behavioral biologists may begin as a source of gathering useful information about your problem (e.g., do birds colliding with windows perceive colors? Are birds attracted or repelled to certain colors?). However, cultivating these relationships may also end up being a source of collaboration. Many of the examples cited in this book started with phone calls or emails between wildlife biologists and behavioral biologists, and evolved into key contributions to the field of conservation behavior.

I am already practicing conservation behavior

If you are a wildlife biologist or behavioral biologist already engaged in conservation behavior, we do thank you for your hard work. It is likely that some of your published work is already cited in previous edited volumes on conservation behavior or in this *Primer* (however, keep in mind that our purpose in writing this book was not to conduct a comprehensive literature review, but rather illustrate applications with examples). You will hopefully find that our *Primer* contains new ideas or methods that you may find useful in solving some of your conservation problems. But, if you have additional novel insights, please, do not hesitate to share them with us so that we can learn more and potentially include them in subsequent editions. As a conservation behavior practitioner, we encourage you to promote the use of the discipline, when appropriate, to your peers, to your bosses, and to the community in general. Given the severity of our biodiversity crisis, we think that the more diverse our toolkit is, the higher the chances of telling success stories.

Adaptive Management:
The *Key* to Conservation Behavior

The best management decisions come from the best science, or so scientists are taught. These days we use "active adaptive management" (Walters & Holling 1990), whereby management plans are modified based on the results of well-designed experiments that collect data on factors or variables that are demonstrably important for conservation or management (e.g., Ministry of Forest and Range 2001). We believe that we must take an empirical approach to demonstrating the utility of integrating behavioral science into wildlife management. Our goal is that in the future, the term conservation behavior will no longer be necessary because behavioral tools will be routinely used and combined with other tools to solve many conservation and management problems. However, we are also very pragmatic. If a behavioral approach does not help solve a problem, it is probably not that useful. After all, properly quantifying and integrating behavior into wildlife management will inevitably add more work for wildlife managers. Time is money. And money not spent on unnecessary work can be used for more necessary tasks or less work by making strategies work better the first time.

For instance, if the goal is to increase breeding success in a captive breeding program, an active adaptive management approach would be to design an experiment whereby one or more factors (such as cage size, diet, duration of daylight, etc.) is manipulated and compared to control groups. Determining what to manipulate would emerge from a good working knowledge of how animals behave in their cages. Thus, simply observing animals and their behavior may provide many insights that can generate formal tests. A second use of active adaptive management might be to increase reintroduction success. If pilot trials suggest that recently introduced animals are killed by predators, or are inefficient hunters, a properly designed adaptive management approach would do something to mitigate predation or starvation risk. For instance, we could provide cover in a "soft-release" pen (Kleiman 1989), provide pre-release predator training (Griffin et al. 2001), introduce animals socially (Shier 2005), or engage in predator harassment to reduce the likelihood that recently introduced animals will encounter predators, and then compare the fate of individuals in such treatments to control groups where no mitigation was employed.

By contrast, "passive adaptive management" occurs when biologists use historical data or data from uncontrolled experiments to come up with "best guess" management recommendations, the fate of which may be studied. It is clear that the inferences made under passive adaptive management are weaker because the approach is either correlational, or the experiments are not properly controlled. Nevertheless, the emphasis is on acquiring knowledge by conducting experiments and collecting data, and this may be the best that can be done under many circumstances.

We believe that it is essential to work within the context of adaptive management, ideally active adaptive management (Blumstein 2007), and that conservation decisions be based on evidence (Sutherland et al. 2004). Expert judgement is important to narrow the context upon which active adaptive management can be applied, because experimental evidence on every behavioral facet is not available and conservationists sometimes cannot wait. We are empiricists; therefore, the application of behavioral knowledge must be both cost-effective and help manage or recover populations. We hope this book demonstrates how we can develop this knowledge and incorporate it into current conservation and management methods.

Examples of How Conservation Behavior Can Solve Wildlife Management and Conservation Problems

Of the many areas where animal behavior can contribute to management and conservation, we emphasize four throughout the *Primer*: improving captive breeding success, improving the success of translocations and reintroductions, managing anthropogenic impacts on wildlife, and managing wildlife in urbanizing environments. We provide a brief introduction to each of these themes.

Captive breeding

The IUCN recommends that captive breeding programs be started when it is likely that a population will become critically endangered or extinct (IUCN 2002). Captive breeding requires that animals are brought into captivity and managed in a way to increase the population size (in captivity) while retaining genetic variation. Ideally, captive breeding is integrated with a re-introduction program (see below). Captive breeding is a difficult and expensive proposition, but it is often required once a species is listed under the Endangered Species Act in the United States of America (**Box 1.2**).

Black-footed ferrets (*Mustela nigripes*) almost went extinct. Individuals from the last-known, relict population were brought into captivity in 1985 to begin a captive-breeding reintroduction program. Unfortunately, canine distemper caused the wild and captive populations to crash. By 1987, the population had dropped to 18 ferrets, so all known ferrets were brought into captivity. Despite these setbacks, captive breeding and reintroduction have been moderately successful so far. The current black-footed ferret recovery plan (USFWS 2006) expects to spend at least an additional $12 million for captive breeding alone in order to recover the species to a level at which it can be de-listed by 2030. In total, an estimated $72 million is required to de-list the species.

BOX 1.2 The United States Endangered Species Act

Initially developed in 1966, the US Congress strengthened the Endangered Species Act (ESA) in 1973. Any species in the US as well as in other countries can be listed. Listing and subsequent management is the responsibility of the US Fish and Wildlife Service. Once listed, all Federal agencies are required to develop recovery plans to conserve threatened or endangered species and are prevented from engaging in activities that would harm such species. Importantly, critical habitat gets designated and protected, and landowners are prevented from killing or injuring individuals of a listed species. Subsequent amendments in 1978, 1982 and 1988 have modified, but not substantially changed, the ESA (www.fws.gov/endangered/esasum.html).

Once a species is listed, the USFWS is supposed to develop a *recovery plan*, which is a plan to recover a species to the point of "de-listing." Many species, though listed, do not have formal recovery plans. Some species have been successfully de-listed, but de-listing often takes many years, and the controversy over the wolves in Yellowstone National Park (www.yellowstonenationalpark.com/wolves.htm) illustrates that de-listing may be more of a political decision than a biological one. The politics of conservation are essential to understand, but are beyond the scope of this book.

In addition to the ESA there are a number of other national and international laws, conventions, and programs to conserve biodiversity (e.g., The Convention on International Trade in Endangered Species of Wild Fauna and Flora [CITES], the Convention on Biological Diversity [CBD], and the Marine Mammal Protection Act [MMPA]. See Groom et al. [2006] pages 104–108 for a discussion of these and others).

California condors (*Gymnogyps californianus*) almost went extinct. Urbanization, conflicts with ranchers, and the use of lead shot by hunters reduced the 1987 population to 22 individuals—all in captivity. An aggressive captive-breeding reintroduction program was started in 1987, and by 2005, the population rose to almost 300 birds, with more than 125 reintroduced to the wild. In the past 20 years, the California condor recovery program has cost between $35–40 million (www.fws.gov/hoppermountain/cacondor/FAQ.html). The Condor population has not been de-listed and de-listing will cost many more million dollars (e.g., to further increase the population size, cover legal costs, etc.).

These are two examples of captive-breeding reintroduction programs that underscore their extremely expensive costs. Fundamental knowledge from behavioral biology can help improve the successes of both captive breeding and reintroduction programs, given the costs of intensive recovery. In some cases, greater behavioral knowledge may help alleviate concerns people have with a particular conservation solution (e.g., reintroduction). These concerns could be about how animals may fare upon release. The Colorado lynx (*Lynx canadensis*) introduction attracted considerable criticism because lynx starved upon release (Bekoff 1999). The red wolf (*Canis rufus*) recovery program provides another example where being able to reintroduce animals hinged on public perception. In this case, people were concerned about the animals' ability to avoid getting killed by cars and there were concerns about the welfare of the animals before and during the reintroduction process (USFWS 2007). In both cases, studies of animal behavior provide information vital for planning and successfully executing a reintroduction.

For a given species, there are many husbandry considerations that must be identified (Kleiman et al. 1997). For instance: How should animals be housed—alone or socially? How much space is required? What should the temperature be in the captive environment? What should the light cycle be? What sorts of food are required? Should any of these parameters vary seasonally? While it may be possible to generalize from what is known from species' close relatives, species may also differ substantially in adaptive ways.

Understanding the unique adaptations that a species has might influence husbandry. For instance, some species show sexually-selected infanticide, a behavior where a male moves into a females home range or takes over another males' harem, kills the offspring sired by the previous male so as to induce the females to cycle sooner and therefore allows him to reproduce rather than caring for the offspring of another male (Ebensperger & Blumstein 2007). If a captively bred species engages in this behavior, it would not be prudent to move males around while females have potentially vulnerable young (Anthony & Blumstein 2000). Thus, a fundamental knowledge of natural behavior may shed light on these and other factors and, by doing so, we may breed animals more efficiently in captivity.

Translocation and reintroduction

When a species or population becomes extinct, recovery depends on taking animals either from another wild location and translocating them to the area for recovery, or taking animals from captivity and reintroducing them to the wild (Kleiman 1989). Unfortunately, many reintroductions fail (Wolf et al. 1996, 1998). This is both an ethical issue (a failed reintroduction means that animals have died; Bekoff 1999, 2002), and a management one (it is expensive to rear up and release animals for them only to die). Nonetheless, translocation and reintroductions remain important tools in the recovery of threatened or endangered populations or species (Seddon et al. 2007), and they may be relevant if we are to restore ecosystem function in communities in which species have been lost (e.g., Smith et al. 2004).

Another use of translocations is to remove problem animals (Linnell et al. 1997; Conover 2002). Typically, these focus on carnivores (wolves and bears), but many homeowners trap and translocate "problem" possums, squirrels, and raccoons. There may also be ethical issues in these translocations if translocated animals die (as many do) because they are moved into unfamiliar habitat.

As we discuss in Chapter 7, applying knowledge of animal behavior may have profound effects on the success of reintroductions. For instance, if animals live socially, they may benefit from being introduced in their social groups (Shier 2005). Even animals that aggregate to reduce predation risk, but do not form complex social relationships, might benefit from being introduced socially (Blumstein 2000).

Captive breeding has risks. Among them are the risks to individuals who do not develop in their natural environment. Many species learn about their predators through experience living in their natural environment. If individuals are reared in a predator-free environment and then reintroduced into a predator-rich environment, it is no wonder that many die. Pre-exposing prey to their predators in such a way that they can acquire experience with them prior to their release into the wild may be an important conservation behavior management tool (Griffin et al. 2000).

Anthropogenic impacts

Recreational activities have been encouraged in recent decades as a way of connecting humans with the natural environment, encouraging local economies, and supporting environmental education efforts. The downside, however, is an increase in the rate of human visitation to pristine areas that could trigger negative effects, particularly when species do not have alternative habitats. Although studying behavioral responses to human disturbance is not the way of establishing whether a species is threatened due to recreational activities, it can provide insights into the mechanisms underlying human-wildlife interactions by analyzing them with the theoretical context of anti-predator behavior. The assumption, empirically corroborated, is that wildlife react to humans in similar ways as they do to predators. Mechanisms explaining anti-predator behavior (e.g., the risk-disturbance hypothesis [Frid & Dill 2002]) can now help us predict the outcome of human-wildlife encounters. For instance, species responses to recreationists are usually aggravated after certain thresholds of visitation, which are likely to be species specific. Understanding these relationships can help us manage recreational activities that focus on wildlife viewing, without eroding biodiversity. In addition, this understanding helps reduce cases of conflicts or incidents affecting human health and safety. Clearly, these benefits can address some of the political components of conservation.

Urbanization

More than 50% of the human population now lives in urbanized environments. This creates environmental problems within cities for species that live on remnant fragments of suitable habitat. For instance, in Southern

California native chaparral birds have higher chances of surviving in large patches where coyotes are present because the abundance of domestic cats decreases due to coyote predation (Crooks & Soulé 1999). In smaller remnant patches, coyotes tend to be absent, which thanks to the overabundance of cats associated with humans, increases local predation on native birds. Predatory behavior is the key to regulating interactions in this human dominated habitat. However, the forefront of the urbanization problem lies at the edge of urban sprawl, as new housing developments enhance habitat attrition, fragmenting and restricting the distribution of wildlife. A deep understanding of the behavioral patterns of dispersion, avoidance–attraction to humans, and habitat selection allows us to develop methods of reducing the negative effects of urbanization, while maintaining certain ecological processes.

Questions Conservation Behavior *Cannot* Answer

As Caro noted in the Epilogue to his 1998 edited volume, conservation behavior cannot answer many conservation or wildlife management questions. In part, this is because conservation biology works at a larger scale than single species. For instance, developing strategies for landscape-level habitat protection is not in the realm of conservation behavior (Caro 1998; Buchholz 2007). However, establishing the definition of appropriate (preferred, required) habitat from the species' perspective, and how connected remnants need to be considered with respect to the species' mobility is in the realm of conservation behavior. Managers often need answers quickly, but conservation behavior studies may take a while. However, there are a lot of conservation behavior questions that may be answered more quickly by using the extensive literature on animal behavior that already exists. For those questions that conservation behavior can address, we believe that the toolkit presented in this book may be useful.

Our Approach in This Primer

We adopt a multidisciplinary approach and the Tinbergian approach, which uses insights, approaches, and tools from different levels of behavioral analysis (**Box 1.3**), to solve applied problems in wildlife management. In the following chapters we illustrate how, we believe, applying conservation behavior principles can be productive.

DEFINE A CONSERVATION PROBLEM   Depending upon the nature of the problem, this may be relatively focused: Why do the captive bred offspring of southern white rhinos (*Ceratotherium simum simum*) not reproduce? (Swaisgood et al. 2006); or rather diffuse: What is responsible for the disappearance of reintroduced Vancouver Island marmots? (Bryant & Page 2005). As with all research, the more precisely defined the problem, the easier it is to study it. In the case of the rhinos, they breed well regardless of if they are in the wild,

BOX 1.3 Tinbergen's Four Questions

In 1963, Niko Tinbergen wrote *On the Aims and Methods of Ethology* and proposed what are now referred to as Tinbergen's Four Questions. These questions, based in part on previous suggestions by Ernst Mayr, helped guide behavioral research for the past four decades. These four logically distinctive and mutually exclusive types of questions about causation, development, adaptive utility, and evolutionary history can be profitably applied to any behavioral phenomenon. Importantly, by asking questions at multiple levels of analysis, our knowledge about behavior is enriched. Broadly, a behavioral question can focus on *how* something works, or *why* it is as it is.

Proximate questions are those employed to explain how something works or how it develops. For instance, studies of functional morphology (e.g., which muscles and bones are used when animals perform a certain behavior) tell us how behavior is patterned and its structural basis. Studies of behavioral genetics identify the degree to which genes are responsible for behavior and the exciting new field of genomics identifies those genes. And studies of behavioral endocrinology tell us about hormonal control or regulation of behavior. These three examples illustrate *causal* questions. By answering them, we learn about how behavior works. A logically distinct type of proximate question focuses on the *development* (or ontogeny) of behavior. Ontogenetic questions might ask about the degree to which a particular behavior requires specific individual experiences to be properly performed, and address the time course of development.

Ultimate questions are those employed to explain why we see the diversity of behavior. For instance, studies that focus on the evolution of behavior tell us how or when a particular behavior evolved. They might also tell us how many times a behavior evolved. To do so, evolutionary biologists construct phylogenetic trees (hypotheses about the relationships between species) and then "optimize" (i.e., map) behavioral traits on these trees. A logically distinct type of ultimate question focuses on the *current adaptive utility* of a trait. Only traits that increase the fitness of individuals will evolve or be maintained by natural selection. For instance, if long legs aid in escaping predators, we expect natural selection for leg length and running speed to evolve. Importantly, these four types of questions (or levels of analysis) produce questions that are mutually exclusive only within a level. Consider bird song.

We can ask about the evolutionary history of song learning. Song learning has evolved in parrots, hummingbirds, and passerine birds. Among passerines, it is seen in a broad group called the oscine birds.

We can also ask about the current adaptive utility, or function, of bird song. Male birds may sing to attract females and to defend their territories from other males. In some species, males that sing more songs have more mates and therefore have higher fitness. It would be illogical to suggest that because male birds sing to defend territories (a finding that emerged from the study of the adaptive significance or function of a behavior), song learning has evolved only once (a finding that emerged from the study of the evolutionary history of song). Questions within each of these four levels of analysis are

(continued on next page)

mutually exclusive only with other questions within that level. Thus, song learning could have evolved once, twice, or three times, or bird song may function as a form of intra- or inter-sexual display, but the number of times it evolved does not bear on its function.

We can ask proximate questions about bird song as well. For instance, recent work on the genetics of song has discovered that humans and birds both express the *FoxP1* and *FoxP2* genes. In birds, these genes are specifically expressed during song learning. A set of neurons, called the higher vocal center (HVC), seems to be responsible for the neural control of song learning. In some species, the size of the HVC is correlated with the number of songs they produce, while in other species, the size of the HVC changes seasonally and becomes largest when song learning is required.

Finding evidence that the HVC does not change seasonally has no direct bearing on whether or not *Fox* genes are expressed during song learning. Nor does it directly bear on hypotheses about the evolution of song learning abilities, or about whether or not males that have larger repertoires have higher fitness. Again, these questions are mutually exclusive.

The beauty of taking a Tinbergian approach to studying behavior is that it forces us to examine qualitatively different sorts of questions. By doing so, we generate considerable knowledge about the diversity of behavior. Recognizing these are qualitatively different questions is essential as well to ensure that arguments about explanation are contrasting different hypotheses at the same level of analysis.

on protected ranches, or brought into captivity from the wild. However, the captive-born offspring of rhinos fail to reproduce. Swaisgood and colleagues (Swaisgood et al. 2006; Swaisgood 2007) have systematically studied rhino reproductive endocrinology. They refuted the hypothesis that there is something systematically wrong from an endocrinological perspective with captive-born individuals but did discover that in captivity, rhino females may have uterine infections. Based on a keeper survey they identified no differences between wild-caught and captive-born females in their reproductive behavior. The hypothesis that older females were reproductively suppressing their offspring was refuted. They found differences in the early social development: captive born females are much more social than they are in the wild. Swaisgood and colleagues continue to test hypotheses to nail down the mechanism responsible for reproductive failure of the offspring from captive-born females.

DEFINE QUESTIONS   It is prudent to narrow down questions that can be answered by working at the conservation behavior interface. Wildlife management is a rich and mature discipline in its own right. There are a variety of methods to estimate the size of a population (Williams et al. 2002) and to determine the likelihood of a population persisting over some time

(Beissinger & McCullough 2002). We must identify *specifically what behavioral knowledge* may be important in helping solve a potential problem. This is key to the successful integration of behavior into conservation biology and wildlife management.

DEVELOP A FOCUSED HYPOTHESIS AND MAKE SPECIFIC PREDICTIONS   If we believe that lack of exposure to predators during some critical period results in predator-naïve prey, we must specify the critical period and define precisely what sort of exposure to predators is important (the smells, sights, sounds, or actual experience interacting with a predator).

DETERMINE DEPENDENT AND INDEPENDENT VARIABLES THAT NEED QUANTIFYING OR MANIPULATING   To continue with our predator recognition example, are we looking to quantify changes in vigilance or some specific antipredator escape strategy (e.g., flight initiation distance)? Some species have unique responses to each of their predators; must the response be predator-specific?

IDENTIFY OR DEVELOP SAMPLING TECHNIQUES   We need to employ techniques that allow us to collect data relevant to testing our hypotheses on the species of conservation concern. Behavioral biologists have a long tradition of working with model species (as do conservation biologists; Caro & O'Doherty 1999), but behavior may be species-specific. Behavioral biologists have developed a variety of methods to quantify behavior in systematic ways (Martin & Bateson 2007; Blumstein & Daniel 2007). And, behavioral biologists have developed methods to manipulate the phenotypic expression of a variety of traits (Andersson 1994). The strength of our inferences depends upon the rigor of our methods.

SELECT ANALYTICAL TOOLS   Video processing, event recording software, statistical analysis, etc., enable us to answer our question. Much as population biologists use specific tools to estimate population sizes (e.g., MARK [White & Burnham 1999]), and population geneticists use specific tools to estimate genetic variation and parentage (Kinship and Relatedness [www.gsoftnet.us/GSoft.html]), behavioral biologists use event-recorders and analysis software to quantify behavior (JWatcher [www.jwatcher.ucla.edu]), sociometric programs to define social groups (SOCPROG [myweb.dal.ca/hwhitehe/social.htm]; UCINET [www.analytictech.com/] Pajek [http://vlado.fmf.uni-lj.si/pub/networks/pajek/]), and other software to quantify space use (e.g., Ranges [www.anatrack.com]). Of course the specific tools will depend upon the question to be addressed.

ANSWER THE QUESTION AND APPLY OUR ANSWER TO THE PROBLEM   We believe that such questions should be designed explicitly within an active-adaptive management program. For instance, if we are testing whether pre-release predator training influences survival following reintroduction, first we want to see if there is an effect of training on predator recognition abilities and then we want to see how it influences later survival. In this instance, we must have formal controls where some individuals are not formally trained and their fate compared with those that are trained.

EXPLAIN HOW THE NEWLY GAINED KNOWLEDGE APPLIES TO THE PROBLEM   When publishing the results of our work, we should explicitly address how the new behavioral knowledge that we generated can be *directly* applied to solve the conservation problem at hand. This helps strengthen the necessary integration of behavior and conservation and makes this discipline a source of novel ideas to address specific problems. And, for those that use these tools, it is important to highlight the fact that the tools and approaches are essential for successful conservation and management outcomes.

APPLY RESULTS TO REAL-LIFE PLANS   Because publication does not necessarily lead to useful application, practitioners of conservation behavioral techniques should adapt their published work to actual management plans.

These nine steps assist us in finding solutions to some (not all) conservation problems, in some cases in coordination with other approaches (genetics, community ecology, etc.).

Further Reading

Clemmins & Bucholz (1997), Caro (1998), Gosling & Sutherland (2000) and Festa-Bianchet & Apollonio (2003) are book-length edited volumes on conservation behavior. Caro (2007) and Buchholz's (2007) exchange in *Trends in Ecology and Evolution* makes for stimulating reading. Pullin & Knight (2009) discuss the importance of evidence-based conservation.

2 Why Do Behavioral Mechanisms Matter?

What are behavioral mechanisms and why should we care about them? Behavioral mechanisms can be thought of as *rules* that animals follow. By studying mechanisms, we study proximate causation (i.e., we explain how animals do things). These can be rules about how hormones influence behavior, rules about how temperature influences sex determination, rules about how individuals select mates, rules about food selection, rules about how animals discriminate between signals and the background, or rules about how animals assess the risk of predation. Identifying these rules is essential because they can be used to develop predictive models. Predictive models allow us to understand how populations will respond to anthropogenic change. Once the models are built, rules can be changed to predict different scenarios, a process that makes these predictive models very useful tools. Let's start by thinking about some physiological mechanisms that underlie demographic processes.

Temperature-Dependent Sex Determination

Temperature-dependent sex determination (also called environmental sex determination) is found in a variety of reptiles (Bull 1980). A temperature difference as small as 1–2°C during incubation will influence the resulting sex of the young. In some species, females are produced at lower temperatures, while males are produced at higher temperatures; in other species, the reverse is true. It is easy to envision the consequences of climate change on offspring sex: A systematic increase in temperature can lead to a systematic bias in the sex ratio. By identifying this mechanism of sex determination in a given species, it becomes possible to manage sex ratios by manipulating incubation temperature. In captive-breeding situations this may be essential to produce animals of both sexes.

Sex Ratio Manipulation

Kakapos (*Strigops habropitlus*) are an endangered New Zealand ground par-
rot. Kakapos are bred in captivity, but managers found that they were producing an excess of males. Sex allocation theory (Trivers & Willard 1973)
predicted that in species where male quality is essential for mating success,
females in good body condition would be able to allocate more energy to
producing males. In the case of kakapo, managers wanted to ensure that all
females were in good condition so they initially provided them *ad lib* food
(i.e., as much as they wished to eat). This created an unexpected problem:
offspring sex ratios were highly male biased! Robertson et al. (2006) successfully manipulated the offspring sex ratio by strategically manipulating
breeding female body condition. Managers weighed individuals and measured their body condition and then put them on custom diets to encourage
some to produce females. This technique can be employed in other captive-
breeding programs and is only possible because the managers had a well-
identified mechanism underlying sex determination.

Ethotoxicology

Ethotoxicology, or behavioral toxicology—the use of behavioral assays to
identify toxic chemicals—is a burgeoning field (Parmigiani et al. 1998;
Dell'Omo 2002). There is considerable interest in how endocrine-disrupting
chemicals alter behavior (Clotfelter et al. 2004). Such chemicals are produced specifically to influence the endocrine system (e.g., birth control
pills). Alternatively, endocrine disruption is an unintended consequence of
a lipid-soluble chemical. Lipid-soluble chemicals are particularly dangerous
because they can bioaccumulate in fat and enter the food chain (Clotfelter et
al. 2004). Because the endocrine system is composed of a series of linked
glands, there may be cascading effects when even a single endocrine gland
is influenced. Consequences include sterility, reproductive abnormalities,
changed growth, loss of immune function, modified cognitive abilities, and
direct mortality (Clotfelter et al. 2004). But environmental hazards are not
restricted to endocrine-disrupting chemicals. We have known for years that
heavy metals and chemicals like DDT and PCBs have similar behavioral,
cognitive, and reproductive effects and can persist in the environment for
years.

All of these deleterious consequences are of interest to conservation biologists and wildlife managers. By developing ethotoxicological assays, we
can identify which behaviors are affected by certain chemicals and how
these chemicals influence demographic processes.

A particularly good example of how chemicals influence behavior comes
from a study of the threespined stickleback (*Gasterosteus aculeatus*) (Bell
2004). Because risk-taking can be viewed in the context of life-history theory (animals with limited residual reproductive value should take more risks;
Nonacs & Blumstein 2010), we should expect anything that might influence

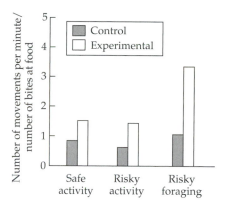

FIGURE 2.1 Adult three-spined sticklebacks were more active and engaged in riskier behaviors when treated with ethinyl estradiol than when treated with a control solution. Graph plots the mean number of movements per minute divided by the number of bites at food. (After Bell 2004.)

reproductive behavior to potentially influence risk-taking and thus mortality. The addition to water of low levels of ethinyl estradiol (an active ingredient of birth control pills and postmenopausal hormone replacement therapy) modified risk-taking behavior in the stickleback (they grew faster and foraged closer to predators; **Figure 2.1**) and increased mortality in experimental females (Bell 2004). Given that estradiol does not break down readily, and that as more women take birth control pills or use hormone replacement therapy, more estradiol enters the water supply, we should expect that wild fish behavior will be altered and that there will be demographic consequences of increased risk-taking. This knowledge may be useful when reintroducing fish or managing the natural (or unnatural) spread of a predator, because increased risk-taking may limit population growth.

Mechanisms of Food Selection

Behavioral ecologists have long been interested in predicting what animals will eat based on the resources available. This is relevant because we can establish the dietary preferences of a population or a species, which can have implications for patterns of habitat use and resource specialization and the degree of competitive interactions between species, which could in turn affect community dynamics. One interesting scenario is to predict the food choice when there is a sharp decline in food abundance. This scenario can occur as a result of habitat destruction that limits the availability of suitable resources in the environment or when individuals are introduced into a new area with different resources from those they had been previously exposed to.

Theory predicts that the breadth of the diet would increase as food resources become less abundant (Roughgarden 1972), given that population abundance is high and that individuals have the capacity (morphologically, behaviorally) to use different foraging strategies. However, there are two mechanisms by which a population can increase their dietary preferences (Tinker et al. 2008): (1) most individuals in the population increase the number of food items they exploit (*within*-individual diversity hypothesis)

FIGURE 2.2 Diet histograms for two sea otter populations (A, Central California; ▶
B, San Nicolas Island in Southern California) and individuals within each of these
populations (C, D). The left y-axis is the proportion of total prey capture events
(filled bars). In A and B, the right y-axis indicates the total niche width (open bars).
In C and D, the right y-axis represents the degree to which the diet of each individual matches that of the population. The x-axis represents the different prey types:
kc, kelp crabs; cc, cancer crabs; cl, clams, scallops, and other infaunal bivalve mol-
lusks; sn, marine snails; sk, small unidentified kelp-dwelling invertebrates; mu,
mussels; ur, urchins; st, sea stars; wo, worms; lo, lobster; ab, abalone; cm, cephalo-
pod molluscs. (After Tinker et al. 2008.)

and/or (2) different individuals specialize in different food items (*among-
individual diversity hypothesis*). Although the overall population outcome
is the same (increase in dietary preferences), these mechanisms can shed
light on whether individuals in the population respond in the same or different ways to a specific change in the environment. The within-individual
diversity hypothesis implies that all individuals respond in more or less the
same way (niche breadth of a given individual should be correlated with
that of the overall population); whereas the among-individual diversity
hypothesis suggests that individuals respond in different ways (niche
breadth of a given individual is a subset of the niche breadth of the population), which supports the idea of individual specialization. An individual
specialist has been defined as "an individual whose niche is substantially
narrower than its population's niche for reasons not attributable to its sex,
age, or...morphological group" (Bolnick et al. 2003).

Sea otter (*Enhydra lutris*) dietary preference has been found to support
the among-individual diversity hypothesis (Tinker et al. 2008). Two populations of sea otters were compared. The population in Central California had
high abundance, but its preferred food had low abundance. The population
in San Nicolas Island (Southern California) had low abundance, but the preferred food had high abundance. Sea otters from the food-poor population
in Central California showed higher dietary diversity, which was the result
of individuals showing food preferences different from those among sea
otters in San Nicolas Island, whose food preferences were similar (**Figure
2.2**). High population abundance may lead to the overexploitation of preferred resources, resulting in different individuals trying different food
sources. This assumes that the costs of learning a new foraging technique
will be lower than the benefits of consuming a new food item.

Establishing the mechanisms of food selection is important to design
strategies for the protection of resources (also see Chapter 6). If the most
common food type of a population is protected but there is a high degree of
individual specialization, key food resources for some individuals may be
lost, which can influence mortality levels. Thus, strategies to protect a wide
range of resources in a population with individual specializations may
buffer the population against environmental changes because this strategy
maintains genetic and phenotypic diversity.

(A) Population average diet (Central California)

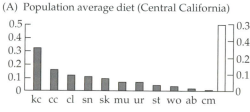

(B) Population average diet (San Nicolas Island)

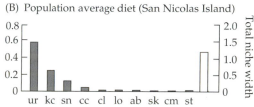

(C) Individual diets (Central California)

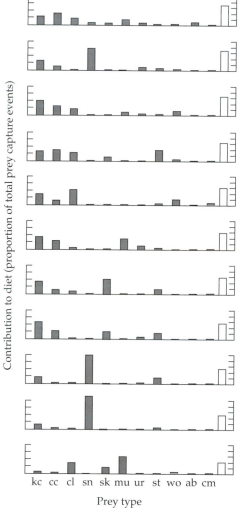

(D) Individual diets (San Nicolas Island)

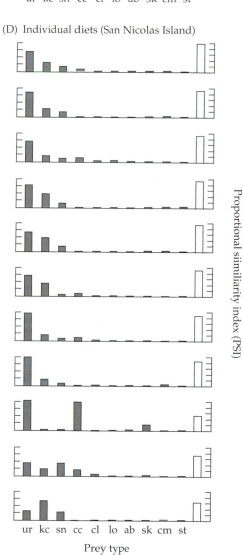

Mechanisms of Predation Risk Assessment

We return regularly to thinking about predation risk because risk-taking behavior influences mortality either directly (animals who take risks may be killed), or indirectly (animals who avoid risks and stay in safe locations may starve). Thus, anything that influences mortality is of interest because differential mortality influences population size. For instance, if "shy" animals hide from predators and have lower predation rates than "bold" animals that not only do not hide but also challenge predators. Let's consider a few ways that knowing the mechanisms animals use to assess predation risk can be useful to managers (see also Chapters 7 and 9).

Humans are increasing pressure on our wildlands. Globally, human visitation to national parks and other protected areas is increasing (Balmford et al. 2009). This magnitude of visitation has consequences for the animals living there. For instance, some species may habituate to humans (i.e., fearful responses decline over time) while others may sensitize to human visitation (i.e., fearful responses increase over time). Preliminary data from Southern California (Blumstein unpublished data) suggest that wetland birds habituate to increased human visitation, while chaparral birds are more likely to sen-

(A)

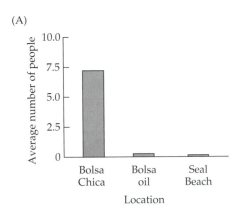

(B)

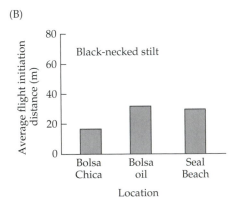
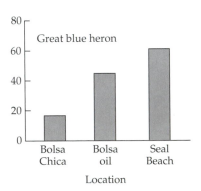

FIGURE 2.3 Wetland birds habituate to human visitation. (A) Number of humans observed visiting three wetland sites: Seal Beach is a naval weapons station with very few human visitors; Bolsa Chica is an ecotourism site; Bolsa oil is a fenced-off part of the Bolsa Chica wetland where oil is extracted. (B) Humans could approach these wetland birds more closely at the popular ecotourism site than at the two areas with fewer visitors, suggesting that the birds had habituated to humans at this site. (After Ikuta & Blumstein 2003.)

sitize. Specifically, flight initiation distance decreases as human visitation increases for wetland birds (Ikuta & Blumstein 2003; **Figure 2.3**), while flight initiation distance increases as human visitation increases for chaparral birds. The reason behind this pattern may be related to the availability of alternative habitats when birds interact with humans. There are very few wetlands in Southern California and all are essential wintering and breeding grounds (Powell & Collier 1998). Birds that need them have no options to move away and avoid humans. Thus, those that persist should to some extent tolerate human visitation (Fernández-Juricic et al. 2009). By contrast, chaparral is widespread (despite coastal development), and presumably individual birds that are disturbed by humans can move away and try to settle in a less disturbed area.

Wildlife managers often implicitly assume that animals that are habituated to people are somehow less "competent" to recognize predators than genuinely "wild" animals. Thus, we might expect habituated animals to have lower predator discrimination abilities. Coleman et al. (2008) tested this hypothesis with Gunther's dik-dik (*Madoqua guntheri*) in Kenya. By comparing predator discrimination abilities with a playback experiment (i.e., animals were exposed to either a predator sound or a control sound) in areas where dik-diks were and were not habituated to humans, they found that habituation indeed affected discrimination abilities but not in the expected manner (**Figure 2.4**). Dik-diks near humans discriminated the sounds of predators from nonpredators. However, dik-diks who lived farther from people and were not obviously habituated to humans

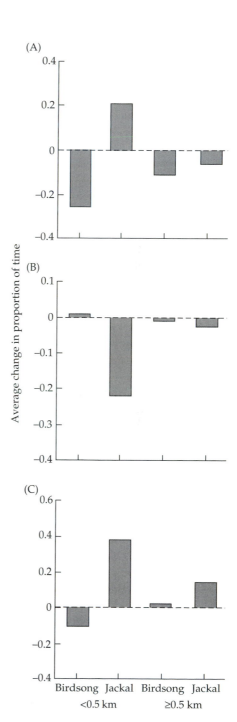

FIGURE 2.4 Dik-dik discrimination of predators (jackal sounds) from nonpredators (birdsong) in areas where they are habituated to humans (<0.5 km from settlements) compared with areas where they are not habituated to humans (≥0.5 km from settlements). Each graph plots the change from baseline in the average proportion of time allocated to (A) vigilance, (B) foraging, and (C) nose twitching. (After Coleman et al. 2008.)

failed to discriminate the sound of predators from nonpredators. This pattern of responses probably occurred because the nonhabituated dik-diks were startled when the experimenters showed up to conduct the experiment, and the lack of discrimination likely reflects a "ceiling effect" (i.e., the animals responded maximally to any sound). The conservation behavior lesson from this experiment is that human visitation can interfere with the ability of animals to discriminate predators from nonpredators and this could lead animals to waste time and energy focusing on humans, rather than "real" threats.

Within the context of predation risk, managers have also been interested in establishing not only when an animal flushes (e.g., flight initiation distance) in response to humans, but also when the animal shows a behavior suggesting the detection of recreationists (e.g., alert distance). For instance, if a bird is looking for food, pecking, and scanning at a certain frequency and when a recreationist approaches to within a certain distance, the animal suddenly increases the amount of time scanning. We can record this response as the alert distance (but the animal has not flushed yet). Alert distances may provide an indication of the distances *beyond* which animals have lower probabilities of changing their foraging behavior in response to human activity, and thus be used to calculate buffer areas (see Chapter 12). However, one of the challenges is to determine the behavioral mechanism by which animals detect humans, and recording such a response under field conditions can be difficult. One possibility is to record changes in cardiac responses, which may occur before animals show alert responses. For instance, heart rates of European starlings (*Sturnus vulgaris*) exposed to humans are more than double their resting heart rates (Nephew et al. 2003).

Yet an alert or cardiac response does not necessarily mean the animal did not visually detect the object sooner but may have delayed its response until the object is close enough that the risks of not tracking it visually outweigh the costs of keeping its foraging behavior. Blackwell et al. (2009a) addressed this issue by establishing the expected distance two bird species—mourning doves (*Zenaida macroura*) and brown-headed cowbirds (*Molothrus ater*)—would potentially detect an approaching object based on their visual acuity and recording the distance at which they reacted behaviorally. Visual acuity was calculated based on eye size and the density of retinal ganglion cells (which take visual information from the retina to the brain). The surprising result is that visually both species were capable of detecting a vehicle approaching from about 1,000 meters away, but they did not react to it until it was about 60–80 meters away (**Figure 2.5**). This finding underscores the importance of assessing the mechanisms of detection as management strategies aimed at minimizing *visual* disturbance can be an order of magnitude away from those aimed at minimizing *behavioral* disturbance.

Mechanisms and Models

Developing the strategies to protect a given species may involve a better knowledge of multiple behaviors rather than a single one (e.g., food selec-

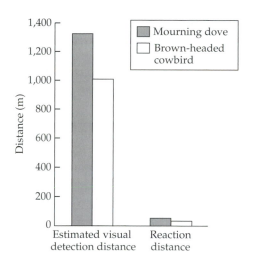

FIGURE 2.5 Visual detection and reaction distances of mourning doves and brown-headed cowbirds. For each species, visual detection distance was estimated on the basis of visual acuity, eye size, and retinal ganglion cell density, and it was given as the distance at which an object 2 meters high could be resolved. Reaction distance was measured in an experiment in which birds were approached by a truck. Animals were in groups of six individuals in a 2.4 × 2.4 × 1.8 meter outdoor flight cage. Reaction distance shown for each species is the average for all individuals in that group. (Data from Blackwell et al. 2009a.)

tion, thermoregulation, acoustic detection). As the number of behavioral parameters to be considered increases, our ability to generate predictions gets more constrained. Individual-based models (IBMs) can be a great tool to address this issue (Sutherland 1996; Grimm & Railsback 2005).

Individual-based modeling follows a bottom-up approach by considering adaptive traits (behaviors with fitness consequences) of individuals as they interact with their environment and other members of the same or different species (competitors, predators), and by assessing the population-level responses (e.g., persistence, changes in abundance) that emerge in time. Thus, IBMs provide a way of investigating the responses that emerge from the various decisions individuals make in response to the environmental variability simulated in a virtual environment.

One of the assumptions is that the behavior of individuals would optimize fitness parameters, which can be modeled directly (e.g., mortality rate) or indirectly (e.g., foraging efficiency). For instance, individuals can choose between habitats with different quality, and if there is individual variability incorporated into the model, they will make a decision not only based on the availability of food in a habitat but also on the number of potential competitors and their dominance status.

The fact that interactions between individuals and between them and the environment can be modeled means that IBMs can be used in conservation biology and wildlife management to generate system-based predictions that follow a mechanistic approach. There are excellent books and book chapters that consider the philosophy and methods behind individual-based modeling and how it differs from classic ecological modeling (e.g., Grimm & Railsback 2005; Grimm et al. 2007).

It is important to briefly summarize the basic criteria that make an IBM stand out from a classic ecology model and to realize the breadth of application and limitation to conservation behavior and wildlife management.

First, IBMs are expected to consider some parameters related to growth, development, or reproduction as these are key components that put individual-based decisions in a life history context. Second, IBMs are expected to consider the availability of resources to individuals rather than system-level constraints to population growth (e.g., carrying capacity). Third, IBMs are expected to treat individuals and their interactions as discrete elements and events, respectively, rather than population-level rates; although these rates can be derived afterwards. Fourth, IBMs are expected to consider variability among individuals that belong to a certain age, sex, and class group. This fourth criterion is an elegant way of modeling important processes that underlie interactions among individuals such as personality traits. Models that meet all four criteria are considered IBMs; whereas if they meet less than these four criteria they are considered individual-oriented models.

IBMs start with a series of rules that animals follow. Such rules may include specifics on foraging (how long to eat, how much to eat), patch searching (how to move through the landscape, when to stop), dispersal (when and how to disperse), and anti-predator behavior (how to respond to a threat). **Box 2.1** describes the steps involved in building an IBM.

IBMs have been applied to human–wildlife interactions. For instance, the availability of time or space for oystercatchers (*Haematopus ostralegus*) to feed on the intertidal zone during the winter can greatly influence their foraging success and consequently their chances of avoiding starvation. West et al. (2002) used an IBM to conclude that small but repetitive disturbances can have a much more negative consequence at the population level than fewer but more disruptive disturbances. The mechanism implicated in this relationship is the lack of available time in the intervals between two disturbance events to forage and replenish energies. Additionally, Blumstein et al. (2005) used an IBM to study the effects of hikers and ecotourists on birds in a forest block. The virtual forest had a hiking trail with different numbers of hikers visiting at different time intervals. When birds encountered hikers they would stop foraging and engage in antipredator behavior by moving away from the disturbance. These simulated variables (e.g., alert distance, flight initiation distance, and movement distance) were modeled based on empirically measured values from a set of different species. The model suggested that the distance that simulated birds responded to humans influenced the amount of food consumed. Because foraging influences body condition, and thus reproductive potential, human disturbance could influence fitness. Comparative reviews suggested that large-bodied birds fled humans at a greater distance than smaller-bodied birds (Blumstein et al. 2005; Blumstein 2006a). Thus, we expect that large-bodied birds will be more vulnerable to human disturbance and the individual based modeling approach developed can be used as a management tool for target species. Specifically, once relevant parameters are empirically estimated, it should be possible to evaluate different usage scenarios and predict the consequences of disturbance.

BOX 2.1 Building an Individual-Based Model

Building an IBM requires following a series of steps that have been described in Grimm et al. (2007). We illustrate these steps with a recent IBM model developed to study the effects of human disturbance on the responses of black-crowned night heron (*Nycticorax nycticorax*) nestlings in a breeding colony located in an area that is planning on opening for recreational activities (Bennett et al. in review). The figure illustrates the different components and some of the processes modeled in this example.

1. Pose a clear question to be addressed with the model. What should be the spatial distribution of pathways and bird-watching facilities (e.g., viewing platforms) to minimize disturbance to the black-crowned night heron nestlings? Nestlings are unable to move away from the nest and subsequently are more vulnerable to recreational disturbance in proximity to the nest. This can potentially lead to chick mortality, nest failure, and ultimately reduced breeding success.

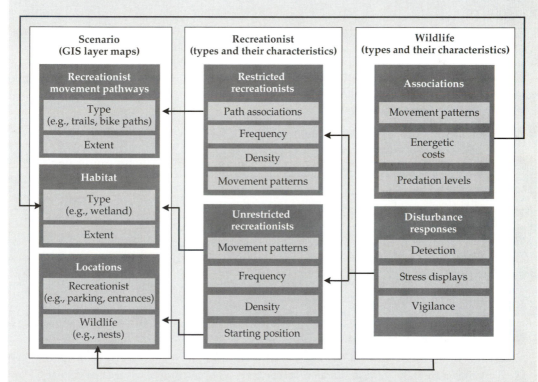

Relationships between the variables and parameters used to define the Simulation of Disturbance Activities (SODA) model structure to study responses of black-crowned night heron nestlings (Bennett et al. in review). The three main categories depicted (scenario, recreationist, and wildlife) define site layout, virtual recreationist activity patterns, and virtual wildlife individuals. Arrows indicate where parameters are influenced by or associated with variables from other categories.

(*continued on next page*)

2. Establish the key processes that govern the system at the individual and environmental level. The responses of black-crowned night heron nestlings depend mostly on the distance between them and recreationists: the closer, the greater the negative effects (Fernández-Juricic et al. 2007b). In the studied system, there are three landscape elements that interact to influence the degree of spatial proximity of disturbance: (1) open water, (2) patches of emergent vegetation (e.g., giant reed *Phragmites australis*) where birds nest, and (3) the upland habitat, used by recreationists but not by herons. The total area of water (therefore water level) ultimately affects the amount of emergent vegetation and the proximity between recreationists and nestlings.

3. Establish patterns that are characteristic of the real system. The model incorporates a modeling space whose parameters have values that are closer and farther away from the real system. It is important to establish *a-priori* the combination of parameter values that reflect the current situation of the system to be modeled in order to compare it with the alternative scenarios. Three water levels (low, intermediate, high) were modeled in the black-crowned night heron study, but the medium level one was chosen as the closest to the real system. Additionally, of all the parameter space (see Figure 2.6), the one without a bird-watching facility with minimal levels of disturbance was the closet to the real system.

4. Build the model structure that captures individual and environmental variability and processes that will cause variables to change over time. The model included three elements: (1) habitat, (2) wildlife, and (3) recreationists. The variability in habitat was represented by the water level (low, intermediate, high). The variability in wildlife responses was modeled from empirical observations on the range of variability in distance at which nestlings freeze (when the nestlings become aware of a recreationist), display agitation (when a recreationist is perceived as being too close), and begin scanning (after a recreationist has moved away). The variability in recreationists was modeled as differences in the spatial position and presence of different types of bird-watching facilities (none, platform with and without vegetation, observation tower).

5. Implement the model. The model called Simulation of Disturbance Activities was written in C++ to track the fates of individuals relative to the behavioral decisions and the corresponding changes in physiological status made by each animal and its environment. Simulation of Disturbance Activities uses site-specific maps created in ArcGIS to define the spatial limits of the virtual elements.

6. Analyze the model. The model assessed how variations in habitat and recreationists affected the frequency of the three different nestling responses, which were analyzed with a univariate and multivariate analysis of variance. A series of sensitivity analyses established whether the application of nestling response parameters at their upper and lower bounds significantly influenced the amount of time nestlings exhibited disturbance-related behavior.

7. Determine the model-stopping rule. The length of the simulation was 30 days, with a 5-minute time step length.

This model found that a combination of management options rather than a single one caused the least amount of disturbance to black-crowned night heron nestlings both temporally and spatially (including the presence, structure and position of bird-watching facilities, presence of different types of recreationist, and extent of suitable habitat). The model also established fine-scale spatial and threshold levels of disturbance (e.g., a bird-watching facility to the north of the core nesting area reduced nestling disturbance only when associated with a 50 meter buffer from the shoreline and less exposed facilities). This model exercise exemplifies the direct management applications of a mechanistic approach for conservation behavior.

Further Reading

Clotfelter et al. (2004) will introduce you to the literature on eco- and etho-toxicology. Roughgarden (1972) discusses niche width. Newphew et al. (2003) illustrate mechanistic responses to stress. Grimm et al. (2007) describe modeling adaptive behavior. West et al. (2002) nicely describe the oyster-catcher IBM. Online links to many individual based models are summarized at: www.red3d.com/cwr/ibm.html.

3 The Evolution of Behavior and Comparative Studies

Behavior evolves. Some closely related species may have similar behavioral repertoires because of common ancestry, while other species facing similar ecological problems may converge on similar behavioral solutions. By making comparisons across species, we can generate a fundamental understanding of the factors that were responsible for the evolution of behavior, and more importantly, we may be able to make *predictions* about potential responses to environmental change by species for which we have *limited information*.

At another level, studies of the same or sibling species at different locations may also provide data that are useful. Some people assume that they must conduct particular studies de novo at their reserves, parks, or jurisdictions. While there is an interesting behavioral literature on intraspecific variation in behavior (e.g., Lott 1991; Foster & Endler 1999), we should not necessarily assume that a particular focal species behaves uniquely at a particular site. Rather, reviewing the literature, and potentially conducting a comparative analysis, may provide important information relevant for conservation and management in cases in which logistic and economic resources are scarce.

In this chapter, we will discuss how we might use comparative methods to make predictions about a species for which we have limited information. Because we typically have limited data on threatened and endangered species, the conceptual approach we suggest, along with the comparative tools we explain, can be incorporated into the conservation professionals' toolbox.

What Is a Comparative Study?

There are several kinds of comparative studies, and we believe that it is essential to be very clear about the strengths and utility of each type. Understanding their attributes may help you solve a particular problem.

Meta-analysis

Meta-analysis (Rosenthal 1991) is a methodology often used by biomedical researchers but also by ecologists (Osenberg et al. 1997), evolutionary biologists (Coltman & Slate 2003), behavioral ecologists (Møller & Thornhill 1998; Stankowich & Blumstein 2005), and conservation biologists (Fernandez-Duque & Valeggia 1994; Hartley & Hunter 1998). **Meta-analysis** combines the results of previously published studies to provide an overall estimate of the *effect size* of a particular treatment.

Effect size is a measure of the magnitude of an effect. While statistical significance tells us whether a particular relationship is likely, effect size tells us something about its importance. A common measure of effect size is r, the correlation coefficient; the larger the correlation coefficient, the stronger the effect. By tradition (Cohen 1988), effects with $r = 0.1$ are considered small effects, $r = 0.3$ are considered medium effects, and $r > 0.5$ are considered large effects.

There are other ways to estimate the effect size. For instance, when comparing the responses to two treatments or of two groups, we can calculate a d-score:

$$d = \frac{(X_1 - X_2)}{SD}$$

Here X_1 is the mean of the first category or treatment, and X_2 is the mean of the second category or treatment. We assume that the standard deviations (SD) of the two distributions are the same, and we use one of them in the calculation, or we could calculate the "pooled standard deviation." Examining this equation, we can see that the d-score reflects a standardized difference between two groups. Larger d-scores imply a greater difference between groups. As with the correlation coefficient, effects with $d = 0.2$ are considered small effects, $d = 0.5$ are considered medium effects, and $d = 0.8$ are considered large effects.

Meta-analysis is most commonly used in biomedical research to answer such questions as: What is the effect of drug X on reducing cholesterol? It is based on the following assumptions: First, if for instance each of 20 small-sample-size studies has a moderately significant effect, or all have non-significant effects in the same direction, this implies that overall there may be a consistent effect in one direction or the other. Second, if for instance 10 studies have one result and 10 studies have another result, this implies that overall there may be no effect. Meta-analysis is particularly useful in helping to identify subsets of species or of a population that behave consistently. Thus, a particular drug may work well in 50- to 60-year-old women but may have inconsistent effects (or opposite effects) in 20- to 30-year-old men.

To conduct a meta-analysis, you must first develop focused hypotheses and then review the literature. Be sure to include all published studies that have tested a particular hypothesis. The exact details of combining quantitative estimates of effect size are beyond the scope of this book, but two key references are Hunter & Schmidt (1990) and Rosenthal (1991), both of which provide step-by-step instructions. Several computer programs are available to help you conduct these analyses (see Further Reading).

Meta-analysis is a powerful tool in conservation biology, for it tells us something about how generally important a particular trait may be: factors with larger effects may be relatively more important than those with smaller effects. By combining the results of a set of smaller studies that test the same hypothesis, we can increase our ability to detect the significance of factors that may have relatively small effects. In general, the smaller the effect, the larger the sample size required to detect it (to convince yourself of this, have a look at a table of significance values for a correlation coefficient, e.g., Rohlf & Sokal 1994). If several nonsignificant studies all have results that are consistently in the same direction (e.g., all have small but positive correlations), by combining the results from that set of studies, we may detect a significant relationship.

Meta-analysis can also be used to identify sets of species that behave similarly. For instance, Stankowich & Blumstein (2005) studied factors that influence flight initiation distance in a variety of animals (see Chapter 12). They found that while group size was positively related to flight initiation distance in most mammals and birds, it was negatively related to flight initiation distance in fish. From this finding they inferred that fish perceived safety in numbers and aggregated when threatened. By contrast, mammals and birds were able to detect a threat at a greater distance, and therefore they fled at a greater distance.

Why would meta-analysis be useful in conservation behavior? With limited knowledge of how a particular species may behave in response to a particular stimulus or treatment (e.g., how animals respond to skiers or mountain bikers), it might be possible to make a better-educated guess based on the generality of an effect that could be revealed through meta-analysis. Importantly, all conservation problems suffer from limited resources, and meta-analysis can help us prioritize efforts. It would be financially prudent to first address factors that have relatively large effects on survival, reproduction, welfare, or any other parameter of fitness relevance. Once those factors are properly managed, it would be appropriate to focus on factors with smaller effect sizes, if resources are available.

Methods to study trait coevolution in a correlative way

Say that we are interested in estimating the home range size for a relatively unstudied rodent. We know that home range sizes are a function of metabolic needs, so generally, larger rodents should have larger home ranges. Thus, if we compile a data set with species' values for a set of rodents (ideally including close relatives of the species in question) and then regress body mass against home range size, we generate a prediction equation that allows us to estimate the size of the home range for a species for which we only have a measure of body mass: $y = mx + b$ is the prediction equation that

FIGURE 3.1   Illustration of a line fitted to data.

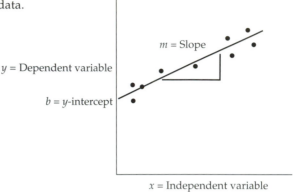

y = Dependent variable

m = Slope

b = y-intercept

x = Independent variable

can be used to predict our dependent variable (y) as a function of the value of our independent variable (x) given the slope (m) and intercept (b) that are estimated from the data (**Figure 3.1**). Such comparative analysis should be a powerful tool in conservation behavior when decisions must be made with limited knowledge about a particular species.

However, a fundamental problem occurs when we do such comparative analyses: a species' trait value (body mass or home range size) is not independent of its relatives. Thus, closely related species have similar values specifically because of common ancestry, and this lack of independence violates a very important assumption of statistical analyses. This lack of independence also has the potential to confound our estimate of home range size (by over- or underestimating values). Because there may be costly consequences of, for example, not protecting a sufficient amount of space or protecting "too much" space, it is essential to properly estimate factors like home range size. To overcome this problem, evolutionary biologists have created several methods that allow us to specifically acknowledge or "remove" this phylogenetic dependence (Harvey & Pagel 1991; Felsenstein 2004; Garland et al. 2005).

All formal comparative methods require a phylogenetic hypothesis. **Phylogenies** (aka phylogenetic trees) are hypotheses about evolution that are created using morphological, behavioral, and/or molecular traits. A phylogeny may simply illustrate the branching pattern of a set of species, or it may include estimates of time since divergence. How phylogenies are constructed is beyond the scope of this *Primer*, but there are several good books discussing their construction (Wiley et al. 1991; Felsenstein 2004). With the explosion of molecular analyses, there are molecular phylogenies available with well-supported estimates of branching patterns as well as estimates of time since divergence. Be aware that there is an ongoing controversy about the ability to estimate branch lengths across disparate taxa, because of different rates of molecular evolution (Graur & Martin 2004; Hedges & Kumar 2004). However, the relative branch lengths within a taxon are less controversial.

There are three main methods to eliminate the phylogenetic noninde-pendence from a set of traits: calculate phylogenetically independent con-trasts, fit generalized least squares models, and use Monte Carlo simula-tions (Garland et al. 2005). The most conceptually straightforward method is to calculate phylogenetically independent contrasts and then fit regres-sions using these contrast values. The elegant logic of independent contrasts is that the difference between sibling species represents independent evolu-tion since these species diverged. Thus, by calculating these difference scores for species, and for reconstructed higher nodes, we generate a data set of $N - 1$ phylogenetically independent data points, where N = the num-ber of species in a data set (**Box 3.1**). Rather than using the species values, we use these contrast values in subsequent analyses.

BOX 3.1 Calculating Phylogenetically Independent Contrast Values

Three things are needed to calculate phylogenetically independent contrasts: (1) a set of traits, (2) a fully resolved phylogeny (to calculate contrasts, we must have a fully bifurcating tree without any polytomies—unresolved phylogenetic relationships), and (3) an underlying model of evolution. There are several pos-sible models of evolution to use. A punctuational model assumes that all evolu-tionary change happens when a species arises; thus, branch lengths on the tree are identical. More realistic models of evolution require estimates of the branch lengths. This can be done a number of ways, such as by assuming Brownian motion (where change is proportional to the length of the branch) Felsenstein 2004; Garland et al. 2005). Selecting the appropriate model of evolution is a non-trivial problem. Often it makes sense to evaluate several models and see whether model choice influences the conclusions (Garland et al. 2005). Free computer programs such as PDAP and PDTREE (http://mesquiteproject.org/pdap_mesquite/), COMPARE (Martins 2003), and CAIC (Purvis & Rambaut 1995) can calculate phylogenetically independent contrasts.

To see how contrasts are calculated, assume the given bifurcating phyloge-ny and four extant species (S1 to S4; **Figure A**). Let's also assume that we have a data set consisting of two traits for each species: body mass and group size. We are interested in predicting group size given an estimate of body mass. To calculate independent contrasts, we first subtract a trait value (e.g., body mass) for species 2 from species 1 (i.e., S2 – S1), and then we subtract the other pair of species (S4 – S3). This creates two contrast values: C1 and C2. A number of methods allow reconstruction of ancestral states (N1 and N2); the simplest is to average the descendants from the node (S1 and S2, S3 and S4), although Brownian motion models of evolution will adjust them according to their branch lengths. With estimates of node 1 and node 2 (N1, N2) we find their dif-ference and create a third contrast value (C3). Because each contrast represents the independent evolution since two species shared a common ancestor, these

(continued on next page)

contrast values are phylogenetically independent, and we would use these contrast values in subsequent regressions. For analysis, the raw contrast values we calculated are divided by the standard deviation of contrast values to create "standardized" contrasts. These standardized contrasts have a mean of 0. Thus, regressions are forced through the origin (see Harvey & Pagel 1991; Nunn & Barton 2001; Garland et al. 2005).

Figure A

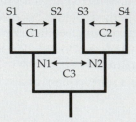

In **Figure B**, we see the result of the regression of contrasts of body mass in relation to contrasts of group size in hystricognath rodents (agoutis, chinchillas, guinea pigs, and their relatives): larger species live in larger groups (Ebensperger & Blumstein 2006). If the slope of this contrast-based regression is not significantly different from the slope of the regression calculated from raw species values, you can use the prediction equation generated by the regression of raw species values to predict group size (Nunn & Barton 2001). If, however, the slope is significantly different, you can use methods to modify the prediction equation generated from raw data. Describing these methods is beyond our scope, but the methods are described in Garland & Ives (2000) and are implemented in the PDTREE program (www.mesquiteproject.org/pdap_mesquite/).

Figure B

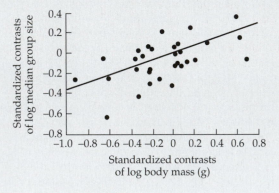

Methods to reconstruct the evolution of a trait

For categorical, dichotomous traits (i.e., those that have one of two values: species present or absent; male or female), it is possible to estimate the trait value of an unknown species as long as you have data on its close relatives. This is a potentially important tool because if we know a lot about a group of less-threatened species but little about their threatened relative, it may be possible to infer the behavior of the rare species based on knowledge of its relatives. There are several methods to do this, and we will discuss one: using parsimony to reconstruct a species trait.

Once you have a phylogeny, you must make some assumptions about how traits evolve. If we assume that traits evolve in a parsimonious way, we can reconstruct the evolution of a trait that has the smallest number of evolutionary transitions (**Box 3.2**). In doing so, we can estimate the value for a species for which we have no data. A potential shortcoming of parsimonious trait reconstruction is that we have no estimate of our error when estimating a trait value. Other techniques (such as maximum likelihood) have been developed to allow us to estimate error; the interested reader can explore these alternative methods in Felsenstein (2004).

BOX 3.2 Parsimonious Trait Reconstruction

Applying parsimony requires us to minimize the number of evolutionary transitions required for the evolution of a set of traits. Given a phylogenetic hypothesis and a set of data that show whether a trait is present or absent in extant species, it is possible to apply parsimony and reconstruct traits. Programs such as MacClade (Maddison & Maddison 2001) and Mesquite (Maddison & Maddison 2005; www.mesquiteproject.org) can conduct these analyses.

In the figure below, we illustrate parsimonious reconstructions given one phylogeny with three different distributions (cases 1, 2, and 3) of extant traits. Black boxes illustrate the presence of a trait in an extant species; white boxes illustrate the absence of a trait. In case 1, a parsimonious reconstruction of this trait suggests that it evolved once, in species C. Indeed, the trait could have evolved in a common ancestor and been lost three times (in species X, A, and B), but that would not be a parsimonious evaluation of the extant evidence. Parsimonious reconstructions are those that minimize the number of evolutionary events. In case 2, it is impossible to generate a single parsimonious reconstruction: the trait either had two evolutionary origins (in species A and C) or one evolutionary origin in the ancestor of A, B, and C, followed by a loss in species B. In case 3, a parsimonious reconstruction suggests that the trait evolved once in the ancestor of species A, B, and C.

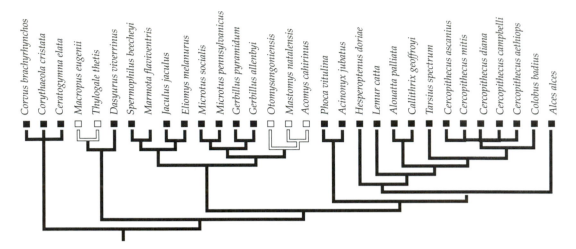

FIGURE 3.2 Parsimonious reconstruction of acoustic predator recognition abilities in a number of birds and mammals (black boxes = demonstrated ability to identify predators; white boxes = failure to identify predators; no box = no data). Black lines illustrate lineages where acoustic predator recognition is predicted to occur. White lines illustrate lineages where acoustic predator recognition is not predicted to occur. Data come from playback studies designed to test acoustic predator discrimination. Note that we initially had no data for yellow-bellied marmots, *Marmota fla-viventris*, a species for which acoustic predator recognition is parsimoniously predicted. Subsequent playback experiments (Blumstein et al. 2008a) demonstrated that yellow-bellied marmots are able to acoustically identify their predators.

As an example, let's consider the evolution of the ability to respond to the sounds that predators make. Understanding the modalities that a species uses to identify its predators is important when we reintroduce animals from captivity, particularly when animals are killed by predators after their release. While many people initially assume that prey should not necessarily respond to the sounds of their predators (after all, predators are quiet when hunting), it turns out that many species have a demonstrated ability to respond to the sounds of their predators.

Blumstein et al. (2008a) collated experimental studies of the ability of a species to respond to the sounds of its predators and asked, given this data set and the phylogeny below: What is the expected ability of a species for which we have no data (yellow-bellied marmot, *Marmota flaviventris*)? Assuming parsimony, Blumstein et al. reconstructed the evolution of the ability to respond to the sounds of predators (**Figure 3.2**).

There are several things illustrated in this figure. First, the ability to respond to the sounds of predators is hypothesized to be an ancestral trait: it is seen both in birds and in numerous mammals. We should therefore expect that various species have the ability to respond to the sounds of their predators, and when preparing animals for reintroduction, we should consider evaluating this ability. Moreover, assuming parsimony, we can predict that yellow-bellied marmots will be able to respond to the sounds of their predators. Playback experiments subsequently verified this prediction.

Conservation Problems

By now you may be thinking: How can these techniques be useful to me? A problem plaguing endangered species management is lack of data on the endangered species themselves. Rare species are difficult to study! There is substantial controversy in the literature about the use of surrogate species to develop knowledge about threatened or endangered species (Caro & O'Doherty 1999; Caro et al. 2005). However, most of the criticisms of the use of a surrogate species focus on using *a single species* as a surrogate. By contrast, these evolutionary approaches use *all available evidence* to estimate traits in a species of interest. We believe that formal comparative methods have the potential to be very useful in conservation biology and wildlife management.

Wildlife biologists may lack data on a particular species diet, home range size, metabolic rate, litter or clutch size, gestation length, estrous synchrony, social behavior, mating system, or "fearfulness." These comparative techniques may help generate "educated guesses" about specific attributes of a given species, genus, or family. There is already a large literature on life history evolution, and a considerable amount of comparative data exists for many species. For instance, *The Birds of North America Online* (http://bna.birds.cornell.edu/bna) and other similar guides for other continents have life history and natural history data for birds. Hayssen et al. (1993) has a lot of detailed life history and behavioral data for mammals, and Rowe (1996) has generally agreed-upon natural history and life history data for primates. Similarly, a good resource for fish is www.fishbase.org.

Wildlife biologists may have to make important decisions about how to house animals. Knowledge of infanticidal behavior and the likelihood that a species will engage in sexually selected infanticide is essential. A number of edited volumes (Hausfater & Hrdy 1984; Parmigiani & vom Saal 1994; Van Schaik & Janson 2000) have been published, and the reviews in them contain data sets that may be useful in predicting whether a species of interest is likely to engage in infanticide. A species' breeding or social system may be able to be estimated from comparative data. Such knowledge will help design housing plans. For instance, it might be particularly stressful to house a socially monogamous species in a mixed-sex cage, or it might be equally stressful to house a colonially breeding species as mated pairs.

Fisher & Owens (2004) suggested that phylogenetic comparative methods could help conservation biologists prioritize conservation research by identifying vulnerable species—those especially prone to extinction. Carnivores are especially vulnerable to extinction, particularly large ones.

Carbone and colleagues used comparative analyses to generate two important findings about mammalian carnivore vulnerability to extinction. In the first study, Carbone & Gittleman (2002) discovered an allometric relationship between the prey biomass and carnivore biomass that exists both within and between species. Thus, given an estimate of prey abundance, they can predict carnivore abundance. Because many species of carnivores are threatened or endangered, maintaining prey biomass is an essential conservation strategy. In the second study, Carbone et al. (2007) quantified mass-related energy budgets (the difference between energy expended while hunting and energy acquired by hunting) for mammalian carnivores. They analyzed small carnivores (those that were < 20 kg) and large carnivores (> 20 kg) separately because their diets were substantially different (smaller carnivores generally fed on invertebrates and small vertebrate prey, while larger carnivores fed on larger vertebrates). In addition to deriving the mass-related energy budgets, which can be used to predict trait values in unstudied species, they discovered that the largest carnivores required large prey. Since large prey are relatively rare, large carnivores are especially vulnerable to extinction.

Such broadscale analyses have also been used to study space use. Carbone et al. (2005) examined the daily distance that individuals of 200 mammalian species traveled (**Figure 3.3**). In a series of analyses, they found that day ranges were predicted by body mass, group size (most prominently in carnivores), and diet. There were also effects of taxonomic order. Their data set is available as an electronic appendix to their paper and can be used to help predict space use requirements for species for which there are limited data.

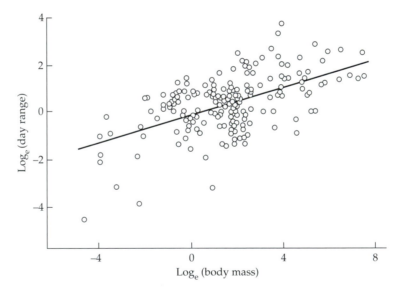

FIGURE 3.3   Bigger mammals use more area each day. Therefore, to protect bigger mammal species, we have to ensure they have access to more space. (After Carbone et al. 2005.)

Further Reading

Felsenstein (2004) is the single most comprehensive resource for constructing phylogenies. Harvey & Pagel (1991) is an excellent book-length introduction to comparative methods. Nunn & Barton (2001) and Garland et al. (2005) are excellent paper-length introductions to comparative methods. Garland & Ives (2000) describes how to back calculate a contrast-based regression onto raw data.

MacClade (Maddison & Maddison 2001) is a Macintosh-based program to study the evolution of categorical traits. Mesquite (www.mesquiteproject. org) can reconstruct traits given a phylogeny and calculate independent contrasts. COMPARE (Martins 2003, www.indiana.edu/~martinsl/compare/), CAIC (Purvis & Raumbat 1995, www.bio.ic.ac.uk/evolve/software/caic/), and PDAP and PDTREE (Midford et al. 2005, available as part of Mesquite) are freely available programs to calculate independent contrasts. PDTREE implements the algorithms of Garland & Ives (2000).

Meta-Analysis 5.3 (http://userpage.fu-berlin.de/health/meta_e.htm) is a free meta-analysis program, and MetaWin (www.metawinsoft.com/) is a popular commercial program. Borenstein et al. (2009) is a book-length guide to meta-analysis.

4 Assessing Food, Habitat, and Mate Preferences

Animals, like us, have specific preferences: they may prefer certain habitat types, food, or mates and may seek or avoid certain stimuli. How do we determine what a species needs and wants? Behavioral biologists have developed a number of specific tools to gain insight into animals' needs and preferences.

There are many situations when we must know something about animals' needs. For instance, we may want to improve a captive-breeding facility. How do we know whether specific improvements are pertinent to the animals? An important lesson from ethology is that animals have their own perceptual world (or *Umwelt*; von Uexküll 1934) that may differ dramatically from our own. Just because something looks bad to us does not mean that it is not sufficient for a given species, and vice versa. We will introduce two concepts used to quantify preferences: price elasticity of demand, and measuring preferences directly using choice tests.

Price Elasticity of Demand

Price elasticity is an economic term for a measurement of the change in demand as a proportion of the change in cost. It is typically calculated by regressing price against demand in a log–log analysis (**Figure 4.1**). But what does this mean? In economic terms, high price elasticity means that as the price goes up, consumers buy less, and when the price goes down, they buy more. By contrast, low price elasticity implies that price has limited influence on demand.

Important things have low price elasticity; they are fundamentally needed, and thus they are insensitive to the amount of effort required to get to them. For instance, if you have a limited income, a gourmet latte is something that might be expendable if the price goes up. Thus, a gourmet latte should have a relatively

FIGURE 4.1   Illustration of the calculation of price elasticity of demand. The slope of a log (cost) × log (demand) regression is the price elasticity. Steeper slopes illustrate higher price elasticity because for any given change in cost, there is a greater change in demand.

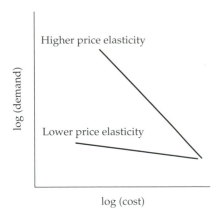

higher price elasticity. By contrast, fuel for your car might have a lower price elasticity compared to a gourmet coffee. After all, you still must get to school or work, and if there is no public transportation, you may have to drive.

This contrast, between fuel for your car and a gourmet coffee, illustrates something important: price elasticity is explicitly a comparative metric. Sufficiently high gas prices will ultimately reduce the demand for gasoline (and therefore should stimulate the development of better public transportation systems). When we are estimating price elasticity, we must do so by comparing different things.

Mason and her collaborators translated this economic idea into animal terms (Mason et al. 2001). They asked captive-reared mink about what cage attributes they were most interested in. We might assume that "satisfied" animals need space, or alternative nest sites, or things to play with, or novel objects. Animal welfare researchers think these things are important, to increase the welfare of captive animals. Recall, however, different species have different *Umwelts*, and thus we must come up with ways to ask animals more directly what they are interested in.

What Mason and colleagues did was to give mink options about what they could "work" to access. The idea was that they would work harder (i.e., spend more effort) to obtain highly valued resources. Specifically, the mink got access to a water pool, an alternative nest site, novel objects, a raised platform, toys, a tunnel, or an empty cage. The challenge was to generate a way to make the mink pay something for access to these resources. The researchers did so by making it harder for the mink to access these resources by adding weights (0, 0.25, 0.5, 0.75, 1.0, or 1.25 kg) to doors that provided access to these resources. Then which "resources" the animals visited were recorded.

What Mason and colleagues found was that as the cost increased, things that were not important to the mink were visited less (**Figure 4.2**). Elasticity was calculated as the slope of a log–log plot of cost (i.e., door mass) × num-

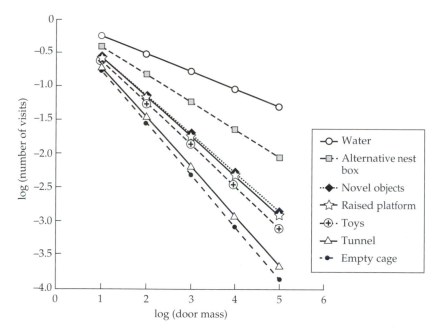

FIGURE 4.2   The relative price elasticity of demand for the mink in the Mason et al. (2001) experiment. As more mass was added to the doors, the change in visitation to water was least effected, while the change in visitation to an empty cage was the most affected.

ber of visits. It was found that the lowest price elasticity was for access to a water pool (slope = 0.26). Following, in order of preference, were access to alternative nest sites (slope = 0.41), access to novel objects (slope = 0.58), access to a raised platform (slope = 0.57), access to toys (slope = 0.62), access to a tunnel (slope = 0.73), and access to an empty cage (slope = 0.77). Thus, the water pool had the lowest price elasticity, and therefore water was the most highly valued resource tested. The conclusion from this welfare experiment was that to make caged mink "happy," we must provide them with water but not be that concerned about extra space, toys, novel objects, or a tunnel.

This is a very promising technique that can be used to ask animals in captivity (and potentially in the wild) how they value various resources we might provide them. Thus, calculating price elasticity is an objective way to quantify value and preference. With thought, it might be possible to apply this technique to study preferences in free-living animals, which are influenced by predators, habitat features, and food resources. Developing such an understanding will allow us to better target management interventions to those preferences that have high price elasticity and, thus, those preferences that will lead to the biggest changes in behavior.

Other Methods to Study Preferences

Animals express their preferences for food, habitat, and mates every day. To create proper captive facilities, we should understand preferences about food and possibly habitat. In some cases, animals fail to breed in captivity. This failure to breed could be caused by a variety of different things, including an inadequate opportunity to select a mate (see Chapter 11). The mate choice literature (Andersson 1994) illustrates two important techniques to experimentally study preferences: simultaneous and sequential choice experiments. To know whether it is likely that animals make simultaneous or sequential choices, we must know how animals aggregate in nature. However, with some knowledge of this, we should be able to design a valid study in captivity to gain insights into preferences.

In a **simultaneous choice experiment**, an individual is provided with two or more stimuli at the same time. Such choice experiments are powerful techniques because the motivational state of the individual being studied is the same when it is exposed to both stimuli. Thus, the inference is that all else is equal and that individuals are able to simultaneously assess all stimuli.

For instance, if we want to know whether a female finch or guppy prefers a male that is more brightly colored, we might place a female in a testing apparatus that has a male on either side of her (**Figure 4.3**). Ideally, the males should not see or otherwise detect each other. The males should differ in color but, ideally, in no other attributes. The assumption commonly made is that a female will spend more time near a male that she prefers. In some cases it is possible to test this assumption by allowing the female to actually mate with a male. A number of experimental issues arise when such a simultaneous choice is allowed.

If you are interested in how important a particular morphological trait (e.g., body size) is, it is important to ensure that the males behave similarly.

FIGURE 4.3   Schematic illustration of a simultaneous mate choice test apparatus. Males A and B are chemically, visually, and acoustically isolated from each other and therefore cannot influence each other's behavior. A female is placed in the center of the middle container, and the time she spends in either side of the container is quantified over a period of time. The performance of specific female behaviors directed at each male is quantified. The behaviors depend upon the species; for example, a female bird may engage in a courtship solicitation display.

If males behave differently (e.g., if more-colorful males court more vigorously), then it is not possible to isolate the effect of color alone on subsequent choice. There are two ways to deal with this. First, quantify male behavior (e.g., courtship intensity), and use this as a covariate in subsequent analyses. Depending upon your specific stimuli, there may be other potentially important covariates that can be quantified. Second, use synthetic rather than real males, if this is possible. For instance, if you are studying whether body size influences female preferences, models or videos of males may be sufficient. If you are interested in quantifying whether males that sing a particular song or make a particular vocalization are preferred, playback of the sounds alone may be sufficient.

Using animations or videos requires a considerable amount of background information on the perceptual abilities of your target species, because peripheral sensory systems are under strong selection and evolve rapidly. Species that must fly or eat living prey may have a higher flicker fusion frequency than more sedentary species (i.e., instead of perceiving the video in continuous motion, they may see a sequence of flickering images). Colors may be perceived differently. Birds have four functional cone types plus ocular media with a wide spectral range, which enables them to see in the ultraviolet spectrum. Generally, the plexiglass that can be bought in home improvement stores does not allow UV light to pass through. Because birds can see in the UV range, this type of plexiglass should not be used in a simultaneous choice test apparatus, because it may block one of the birds' important information channels (some mate choice signals are UV based), and that may bias the final choice. Video monitors, dyes, or paints that look one way to us are inevitably perceived differently by species that have different perceptual abilities.

A few recent studies have proposed another method that can potentially be used in choice experiments: animal robots (**Figure 4.4**; Patricelli et al. 2002; Fernández-Juricic et al. 2006; Partan et al. 2009). Although robots may not be easy to build, recent technology has tended to reduce the size of the elements that cause movement (servos), and that allows generation of more "behaviors" that animals can choose based on visual and acoustic cues. Robots can be built out of skins (which makes them more realistic), from plastic that can be painted, or with added fur, in the case of mammals. Building a robot is as much an art as a science, and it requires a deep understanding of the biology of the model species so that the key behaviors involved in the decision-making process are mimicked. It also requires several iterations of trial and error to ensure the movement simulates a live individual as closely as possible.

Another approach in studying preferences is to use **sequential choice tests**, which present one stimulus at a time to an individual. Whenever an individual is tested on two or more separate occasions, it is possible that something other than the stimulus being tested is responsible for response differences. Thus, it is essential to properly counterbalance such experi-

(A)

(B)

(C)

(D)

FIGURE 4.4   Robotic animals are becoming more common in animal behavior studies. This technology can even be taken to the field to investigate the reactions of wild individuals to the robots. Examples are from four recent studies. (A) A robotic squirrel (photo courtesy of Steven Frischling). (B) A robotic female sage grouse used to study responses of males on a lek (photo courtesy of Gail Patricelli). (C) A robotic California towhee used to study the responses of live individuals to changes in gaze direction (photo courtesy of Esteban Fernández-Juricic). (D) The Falco Robot GBRS (gregarious bird removal system), a radio-controlled aircraft with the size, shape, and color of a hawk, used to repel birds (photo courtesy of Bird Raptor Internacional SL, Spain, http://birdcontrol-birdraptor.com/).

ments. If, for instance, an individual is being tested with two stimuli, the first subject should be exposed to stimulus A first and stimulus B second. The second subject should be exposed to stimulus B first and stimulus A second. The third subject should be exposed to stimulus A first and stimulus B second. And so on. With more than two stimuli, Latin square designs (Winer 1962; Campbell & Stanley 1963) may be employed. When properly counterbalanced, order effects can be controlled, and the response to the specific stimulus type can be identified.

Studying preference in the field is more challenging because we do not have access to controlled environments in which we can reduce the incidence of confounding variables in the behavioral choices. However, field experiment designs and statistical methods to analyze their data have been developed. A must-read book is *Resource Selection by Animals* (Manly et al. 2002), which describes methods for studying resource selection, at different spatial and temporal scales, by animals that use resources in different ways, taking into consideration the limitations of the study (e.g., proportion of available resources known or not, measurements of used or unused resources). Manly and colleagues' methods apply not only to the laboratory but also, and more importantly for conservation and management, to field situations.

One field example on resource selection comes from the idea that in fragmented habitats, relatively large forest fragments have two structurally and functionally distinct areas: edges and interiors. Usually, edges have vegetation that changes from the interior to the open habitat matrix. Edges generate different microclimatic conditions due to greater environmental exposure. Under some specific environmental conditions, edge effects may change species diversity, which may increase at edges because of greater availability of resources (e.g., food closer to the edge, nesting areas at the interior of fragments) but with higher risk of predation and/or parasitism. When these forest patches are surrounded by an urbanized matrix, edges tend to have a higher abundance of human-tolerant species, such as house sparrows (*Passer domesticus*) and domestic pigeons (*Columba livia*), than interior areas (Fernández-Juricic 2001).

Fernández-Juricic and colleagues (2001) asked whether this difference in abundance could affect food selectivity due to intraspecific competition. Changes in food preference could have implications for prey population stability in different areas of a forest fragment, and thus food availability for less abundant species. They focused on birds and manipulated both the density and relative frequency of two types of food to establish whether food selectivity (preferring prey items at low or high relative frequency) would vary between edge and interior areas (**Box 4.1**). They found that selectivity did not vary with food density, but it did with the frequency of food types between edges and interior areas. At the edges, animals did not show selectivity for food items and consumed a greater proportion of the food than at the interior. However, at the interior, animals chose the rare rather than the common food items, which is a pattern that has been observed in other studies (anti-apostatic selection; Allen et al. 1998).

Animals at the edges may not follow this frequency-dependent pattern because of (1) higher intraspecific competition due to higher densities and (2) high numbers of recreationists who disrupt continuous foraging efforts. The implication is that human-tolerant species may affect the penetration of certain plant species into fragments by consuming them differently and thus influencing dispersal and establishment rates of native and nonnative plants.

BOX 4.1 Food Preference Index Used in a Field Test

Fernández-Juricic et al. (2001) studied patterns of prey selectivity in birds living in forest fragments in urbanized landscapes in parks. Three factors were studied simultaneously: location within the forest fragment (edge/interior), frequency of food items, and density of food items. Artificial food items were used (white and brown 1 cm^3 bread cubes) following Allen (1988). In each trial, bread cubes were presented at two frequencies: 10% of one type of bread and 90% of the other. Bread cubes were scattered on a grass lawn in a 1 m^2 area at two densities: 30 cubes/m^2 and 50 cubes/m^2. There were 264 trials total, with 11 trials per treatment (each of the eight location × frequency × density combinations) in each park.

Food selectivity was analyzed with the β index (Manly et al. 2002) by determining the preference for brown prey as $\beta_{brown} = [\log (b/B) / \log (b/B) + \log (w/W)]$, where B and W represent the number of brown and white bread cubes presented, and b and w represent the number of brown and white bread cubes left at the end of a trial. β_{brown} ranged from 0 (total rejection of brown prey) to 1 (exclusive preference for brown prey). One of the problems in calculating this index is that the variability in food consumption across trials could be very substantial. To address this issue, recent studies have used mixed-effects proportional hazards models to better estimate mixed and random effects (see Allen & Weale 2005).

Applications

If reintroduction is a viable option for a species of interest, breeding animals in captivity may be an important part of preventing extinction by providing animals that can be used for reintroductions (Bowkett 2009). However, it is difficult to breed species in captivity, and many such captive-breeding attempts fail to maintain a population, in part because of low fertility or low offspring survival (Synder et al. 1996). In some cases, gametes are moved between zoos and captive facilities as part of population genetic management (Foose & Ballou 1988). Yet such assisted reproductive technology is not guaranteed to be successful and often requires that a number of different procedures (e.g., sperm cryopreservation and genetic sexing of gametes, estrous synchronization, superovulation, artificial insemination, embryo transfer, and in vitro fertilization) be optimized for a given species (Bainbridge & Jabbour 1998). Housing animals socially may be required for successful breeding (Hearne et al. 1996; Swaisgood et al. 2006).

Importantly, however, females may exert preferences (see Chapter 11), and the opportunity to exert them may be an essential factor in mating success. If females preferentially mate with the most dominant animal, even in captivity, such limited mating reduces the effective population size (see Chapter 10; Swaisgood 2007). Of course, the dominant animal can be re-

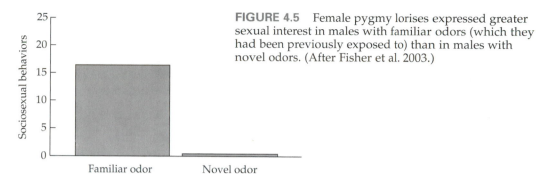

FIGURE 4.5   Female pygmy lorises expressed greater sexual interest in males with familiar odors (which they had been previously exposed to) than in males with novel odors. (After Fisher et al. 2003.)

moved at the time of mating, but this may create a later opportunity for infanticide. For these reasons there have also been attempts to encourage females to preferentially mate with particular males. In both pygmy lorises (*Nycticebus pygmaeus;* Fisher et al. 2003) and harvest mice (*Micromys minutus;* Roberts & Gosling 2004; **Figure 4.5**), females "primed" with the smell of a target male were subsequently more likely to mate with that now-familiar male.

The preferences of many species are quantified to enrich their captive living environments. The best of these studies (e.g., Swaisgood et al. 2001) systematically test enrichment stimuli against controls. For instance, giant pandas (*Ailuropoda melanoleuca*) were provided plastic objects, a burlap sack with straw, tree branches, fruit frozen inside an ice block, and a puzzle feeder. All of these enrichment stimuli modified panda behavior in a way that was considered beneficial for the pandas. Pandas were more active, and expressed more behaviors, while feeding-anticipation behaviors were suppressed, as were stereotopies (repeated behaviors, such as pacing, fence walking, hair pulling, etc., that indicate stress in captivity). Having a diversity of enrichment techniques is important because animals may habituate to one or a few enrichment stimuli.

The approaches we described to measure preferences can fit very well in a captive scenario (e.g., a zoo). However, most wildlife biologists will likely want to find answers to preference questions in the wild. What can we do? This is a great opportunity for finding creative ways of translating controlled experiments into wild scenarios. Wildlife biologists are in a unique position to take the lead because they are the ones with experience observing animals in natural situations and developing objects or manipulating conditions that will attract and engage animals in experiments. This is where expert opinion and experimental design can work together best to apply animal behavior tools to get specific answers from wild animals that can impact their protection or management.

There are countless examples of how wildlife biologists have taken other experimental protocols into the wild. For instance, Kuijper et al. (2009) studied the preference of ungulates for forest gaps versus closed forests in temperate forests in Poland. This is an important issue because ungulate

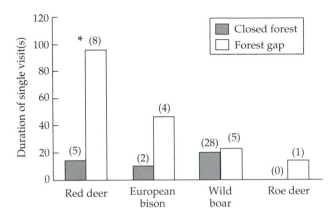

FIGURE 4.6   Mean duration of a single visit to plots of forest gaps and plots of closed forests in four species of European ungulates. Number of visits is given at the top of each bar. The asterisk represents a significant difference between sampling plots. (After Kuijper et al. 2009.)

populations have increased substantially over recent decades, which can lead to negative impacts for forestry practices. When a forest gap is created, higher light conditions increase plant growth rate, which will increase food *availability*. However, plants in a forest gap will invest most of their resources in photosynthesis, which will increase the carbon-to-nitrogen ratio in leaves and twigs, which will decrease food *quality*. To solve the mystery of why ungulates preferred forest gaps, Kuijper and colleagues established six sampling plots in each of two conditions: forest gap and closed forest. To minimize the potential effects of food availability, they planted equal numbers of trees saplings of five different species in each plot, and they measured behavioral activity by recording track plots, taking videos and pictures, and counting dung pellets. They found that ungulates as a group visited more frequently and spent more time in forest gaps in the closed forest; however, this effect was mostly driven by red deer (**Figure 4.6**). The implication of this study is that some forestry practices during clear-cutting after reforestation may enhance ungulate damage due to enhanced browsing behavior. Therefore, population control may need to be complemented by measures that reduce the attractiveness of clear-cuts to ungulates.

Human–wildlife conflict is a huge problem (Conover 2002), and conservation behavior can contribute knowledge and methodologies to its solution. Animals must be kept away from vehicles, crops, industrial enterprises, and homes. A large body of literature describes the use of preference tests that capitalized on either simultaneous choice "cafeteria" trials (e.g., Baker et al. 2005) or two-choice tests (e.g., Campbell & Bullard 1972) to identify effective repellents.

We have illustrated how preferences can be quantified and how they may help us improve the quality of breeding or housing facilities; identify food, habitat, or mate preferences; and study the efficacy of repellents. These are often important questions when animals are bred in captivity and when we are promoting coexistence between humans and wildlife. Thus, these behavioral tools should be in the conceptual toolbox of wildlife biologists working with captive and wild animals. Nevertheless, there is more work to do in integrating experimental techniques into the wild through new approaches (e.g., robotic animals).

Further Reading

Andersson (1994) is a comprehensive book-length review of mate choice. Martin & Bateson (2007) discusses measuring preferences. Critical reviews of two-stimulus simultaneous choice paradigms are discussed in Wagner (1998 and references therein). Manly et al. (2002) provides the experimental design and statistical bases to study resource selection in the field. Resource selection software can be found online (e.g., www.npwrc.usgs.gov/resource/methods/prefer/index.htm).

5 Understanding Habitat Selection for Conservation and Management

Understanding the distribution of animals in space and time is of foremost importance because it facilitates making decisions about which habitats to protect. If individuals are distributed across different patches within a similar habitat type, which patches should be prioritized in terms of protection (e.g., by reducing access for recreational activities, increasing habitat quality, or reducing predator pressure)? The answer to this question depends in a large part on the quality of the patch for a given species. One assumption is that the more individuals found in a patch, the higher its habitat quality. From this, it is commonly assumed that good patches are those where individuals will have high survivorship and reproductive success. This may lead to a heterogeneous distribution (e.g., different numbers of individuals across patches). Thus, by protecting patches with many individuals, one could increase the chance of population persistence over time.

This interpretation, while straightforward, could be misleading. For instance, if our interpretation of high habitat quality is not positively associated with survivorship and reproductive success, then a patch with many individuals may not necessarily be a high-quality patch. Patches where there are many animals that do not do well are referred to as "sinks" (discussed later in this chapter). Establishing the relationship between the number or density of individuals and survival and reproductive success is essential because many monitoring programs use density as a proxy for habitat quality, and areas with more animals are more likely to be protected under this paradigm.

To better predict how individuals of a given species will distribute themselves among habitats, it is also necessary to study the *mechanisms* of habitat and patch selection. As we discussed in Chapter 2, mechanisms are intrinsically associated with behavioral processes. The focus of this chapter will be on how animals decide to establish in a particular patch, whether that decision is successful or not, and the consequences of that decision for the fitness of individuals and the persistence of populations in time. Please note that when we use the word *decide* or *decision*, we are not attributing any special cognitive powers to animals. Rather, we are looking at the choices animals make about what habitat to use and where to settle in that habitat.

Habitat Quality and Abundance

To address the question of whether a species' abundance in a particular area is a good proxy for habitat quality, we need to first examine the mechanism that would lead to such a relationship. The ideal free distribution (Fretwell & Lucas 1970) could be used as evidence of a positive relationship between habitat quality and abundance. In its simplest form, the **ideal free distribution** predicts that animals will establish themselves in patches in proportion to the availability of resources (i.e., patches with more resources will have more individuals; **Figure 5.1**). This relies on several important assumptions (e.g., all animals have similar competitive abilities, patches differ in quality, competition is based on the amount of resources exploited per capita rather than aggressive interactions). Ultimately, however, we expect that individual fitness will be equal between habitats. This assumes that animals have full knowledge of resource availability in all patches and that they move freely between them until they reach the equilibrium (i.e., the number of animals is proportional to resource abundance).

In a seminal paper, Van Horne (1983) proposed that the relationship between local abundance and reproductive success could be negative. If this were so, it would mask an association between abundance and habitat quality. Such a relationship could be produced through an **ideal despotic distribution**, in which individuals that are dominant occupy the highest-quality

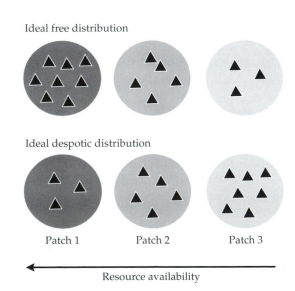

FIGURE 5.1   Number of individuals (triangles) distributed across patches of different habitat quality (circles) under two mechanisms of habitat selection: ideal free distribution and ideal despotic distribution. This simple schematic representation is based on multiple assumptions (see text for details).

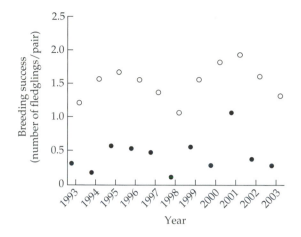

FIGURE 5.2   Breeding success of yellow-legged gulls (*Larus michahellis*) at the Ebro Delta, measured as the mean number of fledglings per nesting pair. Open circles represent vegetated patches; solid circles represent bare patches. (After Oro 2008.)

patches first. Then, as resources become scarcer, dominant individuals force subordinates to occupy patches of lower quality (see Figure 5.1). Thus, the occupation of the high-quality habitat is by a relatively small, but dominant, proportion of the population, and it is lower than the available resources could potentially support. Furthermore, fitness per capita is expected to differ between habitats under the ideal despotic distribution and will be higher in the higher-quality patches. For instance, adult yellow-legged gulls (*Larus michahellis*) prevent younger and more inexperienced individuals from occupying high-quality vegetated patches. Thus, young individuals tend to aggregate in low-quality bare patches, in which clutch size and breeding success are lower because of the effects of seawater flooding caused by strong winds that push water into the flat areas with the nests (**Figure 5.2**; Oro 2008). In general, negative associations between breeding success and local abundance occur in many territorial species (Bock & Jones 2004).

Alternatively, a negative relationship between abundance and reproductive success can stem from abundance that is out of phase with current habitat conditions. Newton (1998) developed a model to explain population regulation in different seasons. Imagine that the peak abundance of a population takes place right after the breeding season, when adults and juveniles aggregate (**Figure 5.3**). During the winter, mortality will reduce population abundance to different degrees depending on individual condition, weather, or interactions between them. If there is little winter mortality, population abundance will be higher than the carrying capacity during the following breeding season, generating a surplus of individuals that may not be able to reproduce because of the lack of territories. This is called a **summer-regulated population**. However, if mortality is high during the winter, because of starvation or hypothermia, the population entering the breeding season will be smaller than the summer carrying capacity. This scenario represents a **winter-regulated population**, which could lead to a negative rela-

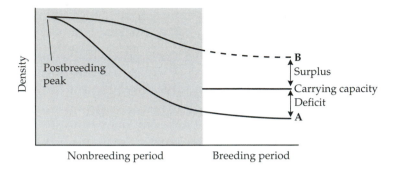

FIGURE 5.3   Possible relationships between seasonal changes in population density as a function of the carrying capacity during the breeding season. A represents a winter-regulated population, where density at the end of the nonbreeding season is lower than the carrying capacity. B represents a summer-regulated population, where density at the end of the nonbreeding season is higher than the carrying capacity. (After Newton 1998.)

tionship (or lack of a relationship) between abundance and habitat quality. For instance, the lower the temperature in the preceding winter, the lower the breeding density of kestrels (*Falco tinnunculus*) in high-quality patches (Village 1990). Winter regulation effects are more pronounced in species with smaller body sizes, because they have fewer energetic reserves as a result of higher surface-area-to-volume ratios (Blem 1990).

Another factor that can generate a negative association between abundance and reproductive success, and thus habitat quality, is human disturbance (Bock & Jones 2004). Recreationists (hikers, bikers, horseback riders, etc.) can reduce the quality of otherwise suitable patches because animals avoid such disturbances (**Box 5.1**). For instance, recreational activities have been shown to decrease use by bald eagles (*Haliaeetus leucocephalus*) of areas with spawned salmon in the Pacific Northwest because bald eagles have a low tolerance of humans (Skagen et al. 1991). Even more interesting is the case of sage sparrows (*Amphispiza belli*), which avoid nesting in habitat patches in human-dominated landscapes (e.g., urban areas) even though breeding success is higher there because of lower nest predation pressure, particularly from snakes (Misenhelter & Rotenberry 2000). This implies that changes in the perception of habitat quality can greatly affect individual fitness, and possibly population abundance. Perception of habitat quality, a critical component in habitat selection, is discussed later in this chapter.

A negative relationship between habitat quality and reproductive success is more pronounced in species with shorter histories of human–wildlife interactions. For instance, this negative relationship is more common in western North America, intermediate in eastern North America, and least common in Europe (Bock & Jones 2004). The implication of a negative rela-

BOX 5.1 Quantifying Human Disturbance in the Context of Habitat Selection

The foraging decisions animals make are influenced by predation risk. Animals trade off the risk of predation with the risk of starvation (Lima & Dill 1990). If human activities create fear in animals, then the decisions animals make about where and how to forage reflect this trade-off (i.e., they will elect to forage in suboptimal locations if by doing so they avoid a perceived risk). By examining foraging behavior and, importantly, by examining the relationship between food consumed and food available, we can make quantitative inferences about anthropogenic disturbances (Gill et al. 1996; Gill & Sutherland 2000). This method assumes that the proportion of a resource used will decline with increased disturbance.

Pink-footed geese (*Anser brachyrhynchus*) feed on agricultural land during the winter. These fields are differentially exposed to traffic, and geese feed farther from roads than they do closer to roads. Gill et al. (1996) found that 50% of the variation in the amount of roots eaten (a favored food) was explained by the distance to the road. They also demonstrated that up to 64% of a favored food (roots) went uneaten near roads.

Gill & Sutherland (2000) thus suggest that disturbance should be quantified by looking at habitat selection at a local scale. At this local scale, it is important to quantify survival, fecundity, and foraging success (which may also be associated with survival and fecundity).

Sutherland (1998a) developed a method to calculate the species-level consequences of disturbance. The key is to calculate the change in population size (ΔN) that results from a specific level of disturbance. He proposed that

$$\Delta N = \frac{LM\gamma d'}{(b' + d')}$$

where L is the total area influenced by some anthropogenic disturbance, M is the density within the site before the disturbance begins (i.e., ideally the density at a patch's carrying capacity), b' is the effect on reproductive output of adding an additional individual (i.e., per capita breeding output), d' is the effect on mortality of adding an additional individual (i.e., per capita density-dependent mortality), and γ is the proportion of individuals that leave a habitat as a result of the disturbance.

The point is that as the population is displaced, by disturbance, from its initial site to other sites, the density will increase at such a refugee site, and this increased density will affect both reproduction and survival of individuals at this site. The ratio of $d'/(b' + d')$ captures the demographic consequence of this displacement, while $LM\gamma$ calculates the number of individuals displaced by disturbance.

Sutherland's model emphasizes the importance of properly quantifying density-dependent effects on both reproduction and survival and that these will have to be known from both breeding and nonbreeding seasons (see text). This is because avoiding people on wintering grounds can have consequences

(continued on next page)

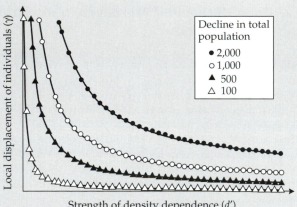

The relationship between density-dependent mortality on the decline in the population and individual disturbance. For any level of density dependence, there will be a greater decline in population size as individuals are locally displaced (i.e., disturbed). (After Gill & Sutherland 2000.)

during the breeding season. As Gill & Sutherland (2000) noted for migratory species, there are few field data on these parameters that are known from both the wintering and breeding grounds, and it is extremely difficult to estimate them. Nonetheless, trying to estimate the strength of density dependence (a nontrivial task that has occupied the entire careers of some ecologists) is essential if this model is to be applied.

It is important to understand the strength of density dependence for another reason: it influences decisions animals make about whether to leave a disturbed location. Strong density-dependent effects will tend to make it profitable for animals to remain in a disturbed area, even if they are negatively impacted there, because by doing so, individuals will avoid the cost of competition in the refugee area.

Often managers would like to use census data to infer something about the relative impact of anthropogenic disturbance. Areas with relatively more animals are presumed to be locations where individuals are doing well. However, the cost of moving from an initial location to a refugee site must be factored into our understanding of the appropriateness of using census data to infer the magnitude of disturbance. Thinking deeply about the effects of density dependence raises questions about inferring the consequences of disturbance by simply noting the number of animals at a particular location (Gill & Sutherland 2000).

tionship between abundance and habitat quality is that we should interpret the results of monitoring programs with caution, in relation to the type of target species and geographic location, to avoid allocating limited conservation resources to areas with low real habitat quality.

Density Dependence Links Behavior and Population Levels of Habitat Selection

Density dependence is the process by which population density has an effect on the per capita rate of population change (r). Density dependence is a framework that has been used to understand how behavioral mechanisms of habitat selection influence population regulation, and it has strong roots in ecological theory (Turchin 1995; Hixon et al. 2002). Behavioral interactions may drive density dependence. Examples of behaviors that drive density dependence include competition for foraging or breeding resources, predators killing a greater proportion of individuals as population size increases, and parasites infecting a greater proportion of individuals as population size increases, to name a few.

Although studies have usually assessed the effects of single density-dependent factors, some species may be limited by two or more factors simultaneously acting at different scales. For instance, black-throated blue warbler (*Dendroica caerulescens*) breeding populations are constrained by two processes: crowding and site dependence, both of which reduce average breeding output (**Figure 5.4**; Rodenhouse et al. 2003). Crowding is regulated by an increase in intraspecific aggressive encounters or interspecific interactions such as aggression and parasitism, both of which increase with local population density. Site dependence occurs when dominant individuals preempt territories. Thus, subdominant individuals are displaced to lower-quality areas and higher-density sites. Site dependence differs from the ideal free distribution in that individuals do not interact directly when occupying patches.

Why is knowledge of density dependence important for conservation behavior? Because it provides strategies that wildlife biologists can use to manipulate populations. For instance, from a theoretical perspective, under an ideal despotic distribution, fitness will vary between patches of different quality. These patch quality differences can lead to source–sink dynamics where surplus individuals in the high-quality habitat (the source) move to the low-quality habitat (the sink). Removing either the source or the sink will influence population dynamics. Removal of sources can lead to local extinction, because population recruitment can be highly constrained. Sinks, on the other

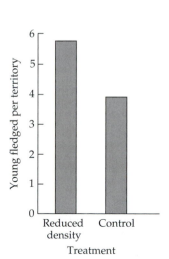

FIGURE 5.4   Mean number of black-throated blue warblers fledged per territory. In the reduced-density treatment, all conspecifics around a randomly selected individual were removed. (After Rodenhouse et al. 2003.)

hand, can play an important role in providing low-quality habitat for young individuals to settle in until, with time, patches open up in the sources. Actually, removal of sinks can reduce the ability of young individuals to disperse and find these vacant new territories, which could lead to an increase in population fluctuations (Morris 2003). Therefore, studying the behavioral interactions between individuals allows us to better predict population dynamics.

One density-dependent model that has been widely used to study patterns of population regulation between habitat types is the **isodar model** (Morris 1995). Assuming that fitness decreases with population density, the isodar model allows us to infer empirically patterns of population regulation based on density estimates between habitats that are qualitatively and/or quantitatively different and adjacent. **Box 5.2** presents an example of how to apply the isodar model with field estimates of population density variations between two habitat types. Establishing patterns of population regulation has implications for assessing (1) habitat selection during natal dispersal, (2) the spatial and temporal effects of disturbance, and (3) the value of different landscape elements based on species' perceptual ranges.

Perception of Habitat Cues in an Uncertain Environment

Many models of habitat selection assume that individuals have complete knowledge of the quality of different habitat patches in the landscape. How can animals gain such information? By exploring different patches. While this assumption is useful in predicting habitat selection patterns in a simplistic way, it is not realistic, because animals would need to invest considerable time in exploring the whole environment to know it comprehensively. There is another complication as well. When an animal is indeed exploring a habitat patch, it is zeroing in on different elements in the environment (e.g., density of brushes, tree cover). Based on these cues, the animal makes a decision but without testing the quality of the habitat directly. If the cues the animal uses to predict habitat quality are misleading, it risks decreasing its fitness.

Why is it important to consider the subtle mechanisms associated with the perception of habitat cues? We could, for instance, argue that individuals using incorrect habitat cues may be selected against. Thus, individuals with the right decision-making process would have higher reproductive success and thus be better represented in the population than those with incorrect habitat assessment algorithms. The problem is that humans are accelerating the rate of environmental change beyond the ability of animals to have a flexible response within a generation (e.g., phenotypic plasticity) or the evolutionary potential to modify their habitat assessment algorithms. A lack of response to rapid environmental change could lead to local extinction. Thus, by studying the behavioral mechanisms behind correct and

BOX 5.2 Using Isodar Theory to Establish Patterns of Density Dependence between Adjacent Habitats

Morris (1987, 1988, 1992, 1995) proposed the isodar model to study patterns of density-dependent habitat selection in relation to changes in population size across landscape elements (Fernández-Juricic 2001). We can imagine two adjacent habitats where there is a decrease in fitness with population density but fitness is higher in habitat A than in B, as in Figure A. Difference in fitness between habitats could affect occupation rates. Habitat A would be occupied first at low densities. But increased density would reduce fitness in habitat A to the level found in habitat B. At that point, individuals could occupy either habitat A or habitat B. We can estimate density in each habitat as the intersection of the fitness–density function with a set of horizontal lines corresponding to equal fitness levels in both habitats (Figure A). An isodar is a line that represents a situation in which the expected reproductive success is the same in both habitats, as in Figure B (Morris 1988, 1995). Changes in the quality of both habitats (e.g., habitats have different sorts of resources) are indicated by changes in the *slope* of the isodar (Figure C). Changes in the quantity of resources (e.g., habitats may have similar resources but in different amounts) are indicated by changes in the *intercept* of the isodar (Figure C).

For instance, habitat selection is independent of density when slopes are nonsignificant. Habitat selection is density dependent when slopes are significant. When slopes are > 1, the density-dependent process is more pronounced in habitat B. Intercepts > 0 demonstrate that initial differences in the overall survival and reproductive success are higher in habitat A than in habitat B. Isodars can be analyzed with geometric linear regression (Morris 1988).

Departures of slopes from unity can be tested with t-tests. We can test whether intercepts differ from 0 by calculating confidence intervals. However, the application of this model is based on some assumptions that need to be assessed in each study system (see details in Morris 1988).

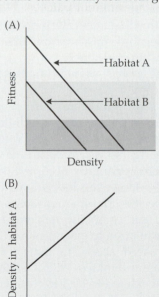

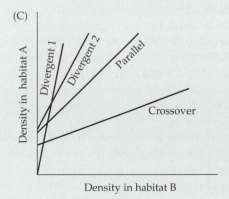

incorrect decisions, we can make better predictions about the quality of habitats for species with different natural and life history strategies. More importantly, understanding the mechanisms of habitat selection can enhance our ability to manipulate habitat quality or habitat cues to attract individuals to the best patches.

Recent research has identified several mechanisms by which animals may respond to cues associated with habitat quality. They are based on three elements: (1) the cues that animals perceive, (2) the cues that are present in a patch, and (3) the quality of the patch measured in terms of fitness parameters (e.g., breeding success, offspring survival, adult survival).

Cues animals perceive

Some habitat cues are preferred over others because animals associate preferred cues with high habitat quality. Individuals can gather habitat cues directly through personal interactions with their environment (i.e., personal information). For instance, red-eyed vireos (*Vireo olivaceus*) use the density of the foliage to settle in woodland habitats, because of the higher abundance of caterpillars in dense foliage (**Figure 5.5**; Marshall & Cooper 2004). However, individuals can also gather cues related to habitat quality by observing the performance of other individuals of the same or different species (i.e., **public information**; Valone 2007). Public information allows for faster and more accurate estimates of habitat quality. For instance, in some colonial species, juveniles and adults that fail to breed in a given year explore different nests at the end of the reproductive season in preparation for the next season (Cadiou et al. 1994). This prospecting behavior is hypothesized to be a way that they can gain information for future breeding attempts. Indeed, kittiwakes (*Rissa tridactyla*) that have failed in their breeding attempts tend to breed in different patches the next season (Danchin et al. 1998).

Another source of public information is the *behavior* of individuals from the same or different species. Individuals can copy the *habitat choices* of conspecifics as indicators of *habitat quality*. The use of public information as a proxy for habitat quality is most likely to occur in inexperienced individuals that lack the discriminatory ability of the more experienced ones (Valone 2007). For example, experienced female budgerigars (*Melopsittacus undulatus*) were less likely to use nest boxes that were apparently occupied than were inexperienced ones (Baltz & Clark 1999). However, individuals can also copy the habitat selection choices of conspecifics for other reasons, such as protection from predators through direct (e.g., mobbing behavior) or indirect (e.g., alarm signals) defense, social stimulation to enhance breeding performance (e.g., high singing rates), increased access to mates, and increased efficacy of defense against competitors (Stamps 1988; Donahue 2006). All the mechanisms we just listed are considered to lead to conspecific attraction. However, we must note that individuals are considered to be using public information only when they copy the habitat selection choices of conspecifics as indicators of habitat quality.

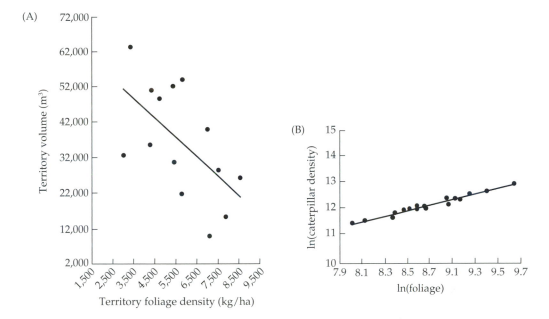

FIGURE 5.5   Red-eyed vireos (*Vireo olivaceus*) appear to use the density of foliage as an indicator of the abundance of caterpillars, an important food item during the breeding season. (A) Territory volume (a three-dimensional representation of a territory) increases with a decrease in the density of the foliage within the territory. (B) The density of caterpillars increases with the density of the foliage during the nesting phase. Foliage density is a stable clue present during the prenesting season and that can predict the abundance of food during a phase in which chicks require an abundant supply of arthropods. (After Marshall & Cooper 2004.)

Conspecific attraction means that individuals are attracted to conspecifics and elect to settle in areas near them. Step back for a moment and consider that conspecific attraction violates a widespread, implicit assumption that individuals will spread themselves out to reduce competition. For species that rely on conspecific attraction, it may be important to translocate animals in groups or to create ways that encourage them to settle together. Conspecific attraction has been identified through both manipulative and correlational approaches in different invertebrate and vertebrate taxa (Campomizzi et al. 2008), and it can explain the patterns of aggregated spatial distribution in many species. For instance, the patterns of grasshopper (*Ligurotettix coquilletti*) habitat selection are influenced by the presence of conspecifics (Muller 1998).

Individuals can also be attracted to settle in habitat patches via **heterospecific attraction**, that is, they are attracted to members of other species. Processes similar to those underlying conspecific attraction can lead to heterospecific attraction. The best example comes from the breeding habitat selection patterns of migratory species in highly seasonal environments (e.g., northern bird breeding communities; Seppänen et al. 2007). Through manipulation of the occurrence of resident *Parus* species, Mönkkönen et al. (1990, 1997) found in both Europe and North America that the density of migratory species was higher in plots where residents were present.

Heterospecific attraction may be particularly likely to occur under certain ecological conditions (Mönkkönen & Forsman 2002). First, given the restricted time available for migratory species to breed, resident species are assumed to be good indicators of habitat quality because of the greater amount of time they have to devote to choosing patches. Second, generalist migratory species are more likely to use heterospecific attraction than specialist migratory species, because the former can use a broader range of cues to make their decisions, whereas the latter are more constrained by specific habitat requirements (generally some component of habitat structure or floristic composition). Third, heterospecific attraction is more likely to occur when resident and migratory species are not competing for the same resources or when resources are not limited while breeding. A recent study found fitness benefits for pied flycatchers (*Ficedula hypoleuca*) that use heterospecifics as cues for settlement. Flycatchers that lay eggs earlier have larger clutches, and the young grow faster when they breed close to titmice (Forsman et al. 2002).

Cues present in a patch

Imagine a pristine grassland environment with patches of different floristic composition. Imagine that certain insects prefer a narrow group of species of grasses. Now, during the breeding season, imagine that a ground-nesting bird tries to find a nesting patch in an area where the grasses are tall enough to prevent easy detection by predators but not so tall that they can outcompete other grassy species that are associated with insects. Also assume that insects are the key resource these ground-nesting birds will use to feed their chicks.

Given this scenario, we expect that when birds find grasses with the preferred height, they will attempt to nest in them (triggering density-dependent effects). Whether an individual nesting in the area of preferred grass height actually is able to survive and reproduce relies on a critical assumption: grass height is a good proxy for a patch with enough food resources (and relatively low nest predation risk) to increase chick survival and fledging success. Individuals that did not secure a patch in the preferred habitat may still settle in patches with less preferred cues (e.g., shorter grasses), but their breeding success may be lower because of higher incidence of nest predation.

In less pristine environments, this story gets complicated because the preferred, or less preferred, cues may be present, or not. Thus, anthropogenic habitat changes may mislead animals' assessment of habitat quality. For instance, areas with the preferred grass height may be sprayed with insecticides, which will greatly reduce the abundance of food resources for the chicks. Another scenario is that in these high-quality areas, the grass may be mowed, which will eliminate the preferred cue, but it could still maintain the necessary floristic composition to ensure presence of insects. Human activities can modify environmental cues. This decoupling of habitat "signaling" from habitat quality may have profound consequences for wildlife.

Habitat quality in fitness terms

Source–sink theory is used by ecologists to predict the fitness consequences of habitat selection. If animals select a low-quality patch (e.g., low resource abundance, high predation), reproductive success will be low even if there are few competitors. Patches with mortality rates higher than birthrates are known as sinks, and they rely on immigration of individuals to sustain a certain population size. High-quality patches (e.g., those with high resource abundance and/or low predation) can increase reproductive success such that the birthrates are higher than the mortality rates. If these high-quality patches also have emigration exceeding immigration, then these patches are referred to as sources. Reproductive success can decrease substantially in source patches when competitor density increases, leading to a potential pseudo-sink. The difference between a true sink and a pseudo-sink is that the latter can become a source with a decrease in the density of breeding individuals, but the true sink condition is irreversible, unless habitat is manipulated.

Putting cues perceived, cues present, and habitat quality together

We can combine the cues animals perceive, the cues present in a patch, and the quality of a habitat and then classify a patch in a human-modified landscape in relation to how animals use it (**Figure 5.6**). When a species' preferred cue is present in a high-quality environment, the patch can be considered a source. On the other hand, when the habitat is low quality, and when the preferred or less preferred cues are absent or when the less preferred cues are present, the patch can be considered a sink. This classification allows us to identify two categories that have received recent theoretical and empirical attention because of their influence in population dynamics (Schlaepfer et al. 2002; Battin 2004; Robertson & Hutto 2006; Gilroy & Sutherland 2007). When habitat quality is high but the preferred or less preferred cues are absent or the less preferred cue is present, the patch is considered an **undervalued resource**. However, when the most preferred cue is present but the habitat is low quality, the patch is considered an **ecological trap**. It is important to search for situations where animals fall prey to ecological traps, because these are areas that can become population sinks.

Undervalued resources can actually be relatively common. For instance, airfields are, for some species, high-quality habitats. This creates a manage-

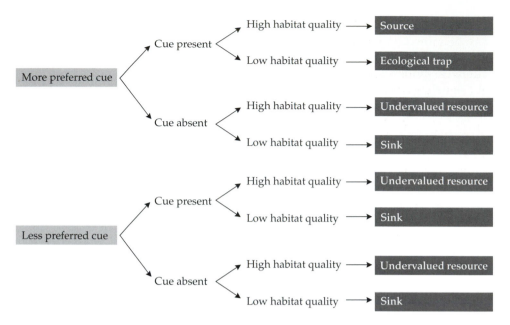

FIGURE 5.6   A schema to classify the different patches present in a human-modified landscape in relation to how animals use them.

ment problem when planes hit birds that are attracted to these airfields. Because of the potential for disasters, there has been a lot of work trying to figure out how to repel birds from airfields, using different means (Blackwell et al. 2009b). For instance, wildlife biologists may employ sound cannons that are triggered to go off at random times, portable laser devices, and even radio-controlled aircraft with the size, shape, and color of a hawk to scare birds and reduce the use of otherwise suitable patches. These patches then become undervalued. Undervalued resources can also be created by removing preferred cues from high-quality habitat, by increasing habitat quality but not switching cues that may attract individuals, or by using less-preferred cues in otherwise high-quality habitat. From a population perspective, a high proportion of high-quality habitats that are underestimated can reduce population size in time because only a few individuals in the population will benefit from the fitness advantage of breeding in an undervalued resource.

Most of the empirical evidence on potential ecological traps comes from avian studies (Battin 2004). For instance, Shochat et al. (2005) found that in tallgrass prairies, grazed plots are ecological traps, compared with unmanaged ones, because birds prefer to nest in them because of the high abundance of insects, but there is also a higher incidence of nest predators (mostly reptiles) that have easier access to nests. Similar examples have been found in areas with high abundance of breeding resources but low breeding success levels due to disease or brood parasitism.

Three mechanisms can trigger an ecological trap (Robertson & Hutto 2006). First, if habitat quality is low and remains unchanged, adding a preferred habitat cue (e.g., by planting more shrubs) can attract individuals that otherwise would not have preferred that patch. Second, if a preferred habitat cue is present but habitat quality is reduced (e.g., by introducing a non-native predator), then individuals will be attracted to a low-quality patch. Third, both cue addition and reduction of habitat quality can act simultaneously. These mechanisms can generate two types of ecological traps. An **equal-preference trap** is made after a cue that was added makes both sources and ecological traps look the same; in this case, animals are expected to settle equally in both habitats. However, besides adding a cue in the ecological trap, if the quality of the cue is enhanced over that in the source habitat, then animals will be more likely to settle in the ecological trap than in the source habitat, leading to a **severe trap**.

Ecological traps are expected to have important population consequences, but sometimes empirical data are difficult to obtain. For instance, ecological traps can lead to population extinction, and the extinction process can be faster in severe traps than equal-preference traps. Small population sizes can also shorten local extinction times if most individuals are in the ecological trap instead of in the source habitat. The proportion of ecological trap patches in a landscape can also affect local extinction times, which can be accelerated with increasingly poor habitat quality in the ecological trap. Several studies have provided ways of improving our ability to detect ecological traps, which will greatly enhance future empirical research (Robertson & Hutto 2006).

Animal Movement

Movement is an intrinsic component of habitat selection. Animals do not stay in one spot after birth and maturation. They move small and large distances (Johnson 1980; Ims 1995). Within a resource patch, animals move very small distances to search for food. Within a habitat, animals move small to medium distances within their home ranges to find resource patches that provide necessary food, shelter, or some quality that makes it beneficial to defend a territory—an area where individuals control access to others. Within a mosaic of the same or different habitat patches, animals will move medium to large distances in a process called dispersal to establish home ranges. Dispersal is particularly important for juveniles that need to prospect for territory. Finally, within a geographical region, animals will move large distances in a process called migration to find better conditions (food, temperature, shelter, etc.) in a given season.

There are excellent books and reviews about how animals move (see Further Reading) that will allow you to gain a better understanding of the environmental, ecological, and social factors involved in movement and, more specifically, of the relationship between movement and habitat selection. Animal movement has been studied from different perspectives. In general, conservation biologists and wildlife managers are interested in the probabilities of animals moving from point A to point B, in order to estimate, among other things, the probabilities of colonization, extinction, immigration, and emigration in different habitat patches in the landscape.

However, the behavior involved in this, that is, *how* animals get from point A to point B, will help us identify important mechanisms that might be suitable for intervention or that might help us identify factors or features that can be manipulated to block or facilitate movement. Thus, we are interested in studying the decisions animals make while moving, particularly when confronted with a habitat type that is different from the one that they have been moving through (**Figure 5.7**). Their decisions will determine a movement path, which could be affected by the species' life history, individual motivation, age, sex, etc. For instance, when forest birds moving through the woods suddenly face a gap of pastureland, will they venture across the gap or will they take a detour to avoid the open area? Or will these birds go through a long fencerow of trees in an open area to reach the other side of the forest? If they do not go through the fencerow of trees, will they settle in one side of the forest habitat with potentially less habitat quality (e.g., higher competitor density that increases intraspecific competition)?

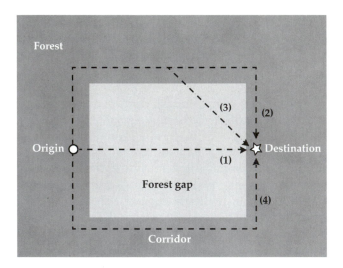

FIGURE 5.7 Decisions animals can make when moving through different landscape elements, which lead to different types of movement strategies. When individuals get to the point of origin, they can reach the destination by (1) crossing the forest gap, (2) taking a full detour through the forest, (3) taking a partial detour that makes them cross an open area but reduces their exposure, and (4) moving through a narrow forested corridor. In playback experiments, the speakers could be at the destination or at both the origin and the destination. (After Desrochers & Hannon 1997 and St. Clair et al. 1998.)

Ultimately, the factors influencing movement paths will determine whether animals coming from point A get to point B or settle for point C. Therefore, our goal is to shed light on the mechanisms influencing the probabilities of movement through the landscape.

Let's think through the behaviors involved in movement. First, animals must reach a degree of motivation high enough to leave the areas they are occupying (e.g., through a decrease in availability of food or territory space). Second, they must cross habitat of the same or a different type or choose alternative routes. Third, they must make decisions about where to stop. Studying these behaviors in the field is quite difficult using observational approaches, because not all the animals we detect will have the same degree of willingness to move. There are different methods of studying movement paths, from color banding to tagging animals with radio transmitters (reviewed in Desrochers et al. 1999; Chetkiewicz et al. 2006), but two of them are particularly relevant if we wish to minimize (or control) variations in motivation: playbacks and translocation experiments (Bélisle 2005). Playback experiments are particularly suitable for studying small-spatial-scale decisions, whereas translocation experiments are appropriate for larger-spatial-scale decisions.

The idea behind playbacks is simple: An animal reaches a point in an area at which calls are played to lure it to a speaker. The animal then needs to decide whether to move to the speaker in a straight line or take a detour, depending on the type of habitat it has to cross and the distance to the speaker. Desrochers and Hannon (1997) presented mobbing calls of black-capped chickadees (*Parus atricapillus*) and red-breasted nuthatches (*Sitta canadensis*) to attract small passerines to move through different-sized gaps of forested habitats during the postfledging season. Combining the responses of the five most abundant species studied, Desrochers and Hannon found that the length of the gap substantially affected the probability of gap crossing (**Figure 5.8A**). Gaps of less than 30 meters decreased the probabilities of crossing very slightly; however, individuals were eight times less likely to cross gaps of 100 meters. Interestingly, animals tended to take detours by going through the forested habitat to reach the speaker (see Figure 5.8B) instead of crossing the gaps directly. In many cases, these detours were twice or three times the distance of the gaps. Bélisle and Desrochers (2002) found that these results are similar in the breeding and nonbreeding seasons, which suggests that forest birds prefer to move through landscape elements that provide some degree of tree cover instead of moving across open areas (e.g., pastures, grasslands). This suggests that in human-dominated landscapes, forest corridors might facilitate movement.

St. Clair et al. (1998) modified the playback technique by having two speakers: one that attracted the bird to a point of origin and another one at a destination point, with speakers separated by 25 to 200 meters. This permitted them to better funnel birds through three specific habitat types: forests, narrow strips of forested corridors, and open gaps. Black-capped chickadees were less likely to travel as the distance to the destination point

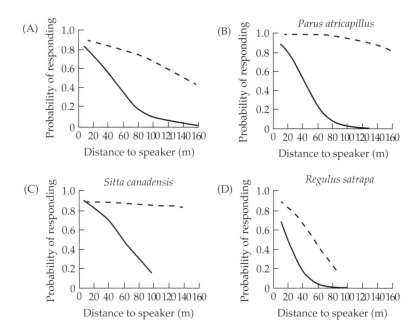

FIGURE 5.8 Probability that smaller forest passerine birds will cross forest gaps of different lengths (distances to the speaker) when presented with playbacks of mobbing calls. Shown are reactions to playbacks conducted across forest gaps (solid lines) and along forest edges (dotted lines). (A) Overall responses of five species. The other panels show responses of three species separately: (B) black-capped chickadee (*Parus atricapillus*), (C) red-breasted nuthatch (*Sitta canadensis*), and (D) golden-crowned kinglet (*Regulus satrapa*). (After Desrochers & Hannon 1997.)

increased; however, this tendency varied with habitat. Chickadees were 4 times more likely to cross 200 meters through forest, and 3.5 times more likely to cross through corridors, than to cross the same distance of open area (**Figure 5.9**). Therefore, forested corridors may facilitate the movement of these small passerine birds over short distances. Using these behavioral experiments to estimate specific probabilities that species will move through different habitat types is a very specific way in which conservation behavior can help landscape planners estimate and (hopefully) increase the suitability of different habitat types when urban development projects take place.

All these playback studies reached an interesting conclusion: responses of forest birds to different gap sizes are highly species-specific (see Figure 5.8B–D). Varying responses can be related to interspecific variations in the perception of risk (see Chapter 7), as some species may feel less safe than others moving through a gap that increases exposure to predators rather than using the protective cover provided by trees. Another potential factor involved is the **perceptual range** of a species: "the distance from which a

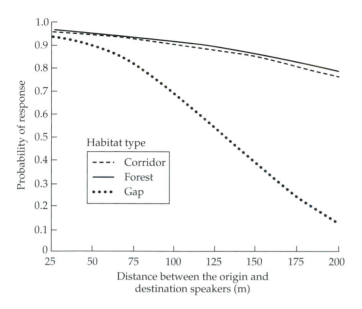

FIGURE 5.9   Black-capped chickadee probability of moving through forest, forest-ed corridors, and open habitat gaps with increasing distance between two points. (After St. Clair et al. 1998.)

particular landscape element can be perceived" (Lima & Zollner 1996). Once reaching the edge of the forest, a species with a large perceptual range will be able to detect the forest at the other side of a forest gap; however, this may not be the case for a species with a small perceptual range (e.g., it may not be able to perceive the other side of the gap). The perceptual range is a function of the time animals have available to search for suitable habitats to move to (Zollner & Lima 2005), the degree of visual obstruction in the habitat that limits the ability to visually detect suitable habitat (Zollner & Lima 1997), and body size (Mech & Zollner 2002). In general, larger species have relatively larger eyes, which provide visual acuity to resolve visual stimuli (e.g., suitable habitat) from farther away (Kiltie 2000). Other factors, such as conspecific attraction (see Chapter 10), can increase the motivation of forest species to move across gaps.

The other technique used to study movement paths, but through larger scales, is temporary translocation. An animal is captured in one location, marked, and moved to another location, and the time it takes to get back to the origin is measured. This technique works particularly well with territorial animals that have high motivation to return to their territories. In translocations, the origin (spot in which the individual was caught) becomes the destination. The direction in which the individual is taken can be chosen to assess the effects of different landscape elements that need to be crossed in the path toward the destination point. For instance, forest birds increase the return time when they have to cross multiple landscape barriers (e.g., rivers, roads, forest gaps) than when they return through the same patch of continuous forest (Bélisle & St. Clair 2001). This suggests that landscape barriers add travel costs by decreasing the speed of movement, as animals' perceptions of risk may increase when they are confronted with visual and acoustic disturbance (e.g., noisy roads), and by potentially increasing energetic expenditure. As somewhat expected, translocation experiments with mammals (Zollner & Lima 1997) have demonstrated that their ability to orient in the landscape is more constrained than that of birds.

Translocation experiments become even more powerful when we can trace the exact movement paths of individuals rather than simply their return times. For instance, Gillies & St. Clair (2008) used radiotelemetry and GPS units to track the movements of translocated individuals from two species in the tropical forest in Costa Rica: barred antshrikes (*Thamnophilus doliatus*) and rufous-naped wrens (*Campylorhynchus rufinucha*), a forest specialist and a generalist, respectively. Barred antshrikes had a greater success of return when using riparian corridors, but they detoured through forest habitats when given the choice of using linear fencerows of individual trees or pastures. By contrast, rufous-naped wrens tended to use fencerows when returning, but return success did not vary between habitat types. Therefore, replacing native forests with pasturelands may increase the resistance of the habitat matrix that forest specialists have to move through. Translocation studies like this that use passive and active remote sensing methods (e.g., sonar, acoustic telemetry tracking, satellite telemetry) are also used to study movement behavior in fish (Lowe & Bray 2006), which has implications for the use of marine reserves.

These types of behavioral studies of animal movement help us establish the **functional connectivity** of a landscape, which is the extent to which a combination of landscape elements enhances or limits the movement between habitat patches (Taylor et al. 1993). Establishing functional connectivity for different taxa is important in managing populations in urbanized and fragmented landscapes, as restrictions in movement through the landscape can affect processes at the individual level (e.g., increasing mortality of individuals that move through open areas) and at the population level (e.g., increasing local extinction of isolated populations that have lower immigration rates).

Conservation Implications

Figure 5.6 provides a framework for thinking about habitat selection when dealing with conservation or management problems. It is likely that managers face complex situations with landscapes that have different proportions of sources, sinks, undervalued resources, and ecological traps. These proportions are likely to vary from species to species. For instance, larger species with wide home ranges may create more management challenges, because of the spatial scale over which their foraging and breeding behaviors take place, than smaller species with small home ranges. Therefore, managing on a case-by-case basis is important, particularly because habitat cues tend to be species-specific. However, improving habitat for a resident species may also positively affect migratory species that rely on heterospecific attraction.

Several conclusions emerge from examining Figure 5.6. For instance, the number of scenarios that could lead to undervalued resources or sinks seems higher than the number of scenarios that lead to source habitats or ecological traps. Recently restored habitats are likely to act as undervalued resources unless the addition of preferred cues is part of the restoration process. Underestimation of habitat quality can actually cause a lag in the responses of species to restoration and decrease the apparent "success" of restoration projects. Monitoring programs before and after restoration should adjust for this possibility by adding surveys at different times after the restoration, taking into account the presence/absence of preferred cues.

Although habitat restoration is a useful conservation tool for increasing habitat quality, it is also expensive to implement. One potentially less expensive alternative is to maintain the quality of patches in a landscape and to manipulate habitat cues if a species shows signs of using ecological traps or not using undervalued resources.

For example, consider a case of secondary fragmentation (Soulé et al. 1992). Secondary fragmentation creates an increase in indirect disturbance (e.g., recreational activity) within fragments that does not directly affect habitat structure but rather influences habitat quality through the avoidance behavior of some species. Under this scenario, rerouting recreational activities through low-quality areas may increase settlement patterns of highly sensitive species.

It is important to realize that by manipulating a habitat cue for a target species, we may inadvertently negatively affect other species using similar cues, because the cue manipulated does not translate into high habitat quality for these nontarget species. If a nontarget species is social and relies on public information about habitat quality, the wrong decision of a few individuals can influence settlement decisions of a large proportion of the population through information cascade effects (Giraldeau et al. 2002). Therefore, we can inadvertently increase the risk of local extinction events for nontarget species. It is important to evaluate the likelihood of these outcomes before starting active manipulations.

To manipulate cues, it is necessary to first identify the cues. This calls for basic behavioral research on the cues animals use to settle in breeding or wintering habitats, particularly for threatened and endangered species. This is particularly important when determining whether second-growth forest is as good for wildlife as old-growth forest, and particularly the age at which reforested areas can be used as habitat by species with narrow habitat requirements. If a significant proportion of the species would not use a reforested area, then strategies for habitat manipulations may need to be tweaked. Establishing the specific clues that forest specialists use to settle, which may require studies on preference and visual perception, may shed some light on restoring habitat to enhance the success of target species.

Nevertheless, even knowing the habitat cues that overabundant species use may be useful if we can manipulate cues to exploit their perception of habitat quality and drive them to ecological traps and thus reduce their population size.

Understanding the theory and methods of habitat selection is key to providing informed scenarios that enhance or reduce the incidence of various species. Habitat selection concepts are also very relevant for the design of protected areas and corridors (Kramer & Chapman 1999; Beier et al. 2008; Fraschetti et al. 2009).

Further Reading

Habitat selection has a long research history. Kramer et al. (1997) presents a superb account of basic and advanced concepts on habitat selection, and Johnson (2007) provides an overview of approaches and methodologies to study habitat selection. Bernstein et al. (1991) provides an excellent review of how knowledge on ideal free and ideal despotic distributions can help solve some applied problems. Animal movement references to get started with are Stamps (2001), Bowler & Benton (2005), Chetkiewicz et al. (2006), and Clobert et al. (2008) We strongly encourage reading these papers as supplements to this chapter. Links to software with which to study habitat selection can be found at http://nhsbig.inhs.uiuc.edu/, www.google.com/Top/Science/Biology/Ecology/Software/, and www.humboldt.edu/~mdj6/585/Wildlife%20Software.htm.

6 Understanding Foraging Behavior for Conservation and Management

Animals must consume food to survive and reproduce, and this vital process has attracted the attention of theoreticians and empiricists for decades (Stephens et al. 2007). We should consider the behavioral mechanisms animals employ when making decisions about where, when, and what to eat, because by understanding them, we may be able to strategically manipulate behavior, which may ameliorate some management problems.

Here, we focus on two areas. First, understanding the costs of foraging can enhance our ability to predict and manage where and when individuals will forage. Thus, by varying the costs of foraging, we could modify habitat use and ultimately population density. Second, the constraints in gathering foraging information (e.g., detection failure or enhanced detection) may influence the ability of individuals to consume different types of food resources. By modifying the perception or perceptibility of food, we may attract individuals to or repel them from particular resource patches, which can eventually affect patterns of habitat selection. By manipulating the foraging component of habitat use, we could influence a population trajectory.

Costs of Foraging

Following the classical foraging studies, let's imagine small passerine birds (e.g., great tits, *Parus major*) trying to maximize their rate of food intake during the relatively short winter daylight hours. To make life a bit easier, tits decide to visit bird feeders that are distributed across an agricultural landscape. Tits face three main types of foraging costs: metabolic costs (MC), missed-opportunity costs (MOC), and predation costs (PC) (Brown 1988). Variations in any of these costs can change the number of seeds per unit time that birds can consume (or leave unconsumed) *within* a feeder at a given point in time. More importantly, if these costs vary differently *between* feeders, the number of seeds consumed (or left unconsumed) will change across feeders. In other words, some feeders in the landscape will have lower foraging costs, and birds will eat more seeds from them.

To understand the three different costs, let's consider a hypothetical scenario. First, imagine that all the bird feeders within the tits' home range have the same amount of food. Second, imagine that the feeders are all the same shape and color and are placed in areas with similar microclimatic conditions (temperature, humidity, light intensity), and the seeds are all in the same type of substrate. Third, imagine that the chances of a predator attack and the bird's detecting that attack are the same across feeders (e.g., feeders are placed at the same distance from cover and the same height from the ground). Equalizing each of these costs across all feeders will also equalize the overall costs, and consequently the intake rate should be the same across feeders. The implication is that the probability of using any feeder with be the same across the landscape.

We can test this prediction by measuring the number of seeds that our hypothetical bird consumes. To do so, we must follow our bird from feeder to feeder (or have cameras in each feeder) and record how many seeds it swallows. This measurement can create some logistical problems. A simpler method is to measure the quantity of seeds left uneaten within a feeder. If we know the mass of seeds originally present, we can subtract from that the seeds left, to get an estimate of the *quitting–harvesting rate*. This variable is known in the literature more commonly as the **giving-up density** (and is abbreviated GUD), and it represents the amount of food left unconsumed when the animal leaves the patch as a result of variation in the three foraging costs (Brown & Kotler 2007). Following our hypothetical scenario, we can predict that when all foraging costs are equal, there will not be significant variation in giving-up densities across feeders.

If we now vary the amount of food available in feeders but keep the other two costs constant, then if our bird is foraging from a feeder with few seeds, it could be missing the opportunity (as daylight time during the winter is constrained) to exploit food from a richer feeder. Putting this missed-opportunity cost in different terms, if 99% of the feeders within the individual's home range have very little food, the bird will probably exploit the 1% of feeders that have a lot of food to a greater extent, reducing the number of seeds left and, thus, the giving-up densities. For instance, Olsson et al. (2002) found that European starlings in high-food-density areas consumed less food per patch than individuals in low-food-density areas (**Figure 6.1**). This outcome is probably the result of a reduction in the marginal value of energy in habitats where food is readily available instead of being a limiting factor.

If we keep the amount of food and the predation risk constant and we then vary the microhabitat locations of feeders such that some feeders are

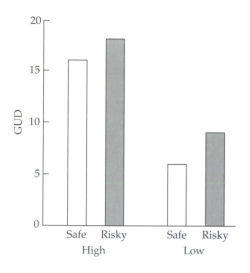

FIGURE 6.1 European starling giving-up densities (GUD) in areas with high and low food availability. Starlings consumed less food (there were higher giving-up densities) in areas with high food density than in areas with low food density, irrespective of the levels of perceived risk (safe vs. risky patches). (After Olsson et al. 2002.)

exposed to windier conditions than others, then we are modifying the metabolic costs of foraging. Animals exposed to higher wind speeds in cool environments experience a reduction in body temperature to below a lower critical temperature, which triggers an increase in metabolic rate and a behavioral response (e.g., shivering) as a compensatory mechanism. Thus, our bird may quit feeding sooner in the feeder that increases the metabolic costs. Kotler et al. (1993) found this relationship between temperature and giving-up densities in gerbils in the Negev desert.

Finally, if we keep the missed-opportunity costs and metabolic costs constant and we vary the distance between the bird feeder and the closest tree that the animal can escape to if an aerial predator (e.g., hawk) attacks, then we are varying the predation costs. In a feeder that is farther away from the closest tree, it will take the bird longer to reach protective cover, giving more time to the predator to capture the individual, increasing the chances of mortality. We can predict that the animal will forage less at feeders in patches with higher predation risk, thereby increasing giving-up densities.

The relationship between predation risk and giving-up density is the one that has received the most attention. One of the reasons is that the costs of predation have been estimated to be much higher than the metabolic costs and missed-opportunity costs of foraging. Brown et al. (1994) estimated that the costs of predation varied between 47% and 91% in gerbils, ground squirrels, and kangaroo rats. This means that animals will be highly sensitive to variation in predation risk and, more importantly, that *giving-up density can be a proxy for change in the perception of risk.* For instance, giving-up densities were found to be higher for desert rodents on nights with owls and moonlight than on nights without them, probably because owls can track prey more easily when there is moonlight (Kotler et al. 1991).

Going back to our bird example and considering all foraging costs varying simultaneously, the question is: When should our bird leave the feeder? Under patch optimization conditions, the theory predicts than an animal should quit when the benefits given by the food energy consumption (H) are no longer higher than the sum of the missed-opportunity (MOC), metabolic (MC), and predation costs (PC), that is, when $H = MOC + MC + PC$ (Brown 1992).

This framework allows us to establish a mechanistic basis to predict when animals will be exploiting foraging resources; we can assess the factors that will affect the costs of foraging and, in turn, the giving-up densities. This is important for conservation purposes because by *manipulating* these factors, we can vary the *suitability* of foraging patches available to animals. For instance, we may be able to reduce the likelihood that animals will forage under dangerous conditions (e.g., close to roads, leading to road kill-sand near runways, leading to bird strikes), or we may be able to increase the chances that animals will use novel food patches (e.g., by attracting birds to forage in safe spots after reintroduction efforts).

Predation Costs of Foraging: A Tool To Monitor and Manipulate Foraging Patch Suitability

The predation costs of foraging described above are relevant in nonpredation situations as well. Let's keep in mind that estimated predation costs represent how animals perceive risk. Many human activities create risks for animals. For instance, imagine a skunk trying to cross a road to get to a food patch. Roads are dangerous, and the skunk could be hit by a car and killed. If it is able to perceive this risk of mortality, then roads increase the predation costs of foraging, compared with habitats without roads.

The predation costs of foraging are an important element in the mechanistic basis of the resource-use–disturbance trade-off hypothesis (see Chapter 12). Let's assume that animals perceive humans as predators. Then, we can imagine that animals might perceive recreationists hiking through a resource patch in a protected area as things to be avoided. This hypothesis predicts that the reduction in the intervals between groups of recreationists passing by may decrease the temporal and spatial availability of resources for animals, and reduce the suitability of the patch. Why? Because frequently disturbed patches will be perceived as having higher predation costs, animals will leave those patches sooner, increasing the giving-up densities.

This disturbance cost can be quantified. Gill et al. (1996) did so by looking at the availability of resources in a patch. They did this (rather than focusing on GUDs) because animals may leave foraging patches simply because the level of resources is so low that it does not pay to stay. Thus, if we have several patches with different levels of human disturbance, we can study the degree of sensitivity of different species by looking at their patterns of food consumption. A species that consumes resources in proportion to their availability can be considered less sensitive to disturbance than a species whose consumption decreases with increasing disturbance and thus leaves food in a high percentage of food-rich patches.

There are other, more subtle forms of disturbance that may influence the costs of predation. For instance, many species of birds and mammals use both vision and hearing to detect predatory threats. Consider a bird foraging by pecking on the ground. Pecking reduces the amount of information about potential risks that the bird can gather using its visual system, because most of the bird's frontal view is focused on the ground, and its peripheral vision is also constrained to a certain degree by the size of the blind area at the rear of the head (Tisdale & Fernández-Juricic 2009). Under such head-down conditions, the bird may rely on its auditory system to detect sounds that can provide cues about potential threats. However, when noise levels increase in amplitude (volume), the auditory pathway is constrained, and thus animals should be more fearful. As a result, individuals foraging in noisy conditions may increase the amount of time they spend head up (reducing head-down time), to compensate for the lack of information (Quinn et al. 2006). Therefore, noise can increase risk perception because animals are less able to detect a threat.

The predation costs of foraging can give managers a tool for identifying the habitats that animals perceive as risky, by recording giving-up densities in the field. The findings can then be used to identify spots that require measures to reduce disturbance (e.g., restricting visitors during times of the year with low food availability), or they can be used as an indication of how animals perceive risk before substantial changes in the landscape (e.g., urbanization). For instance, Bowers & Breland (1996) found that gray squirrel giving-up densities were lower (more food consumption) in suburban than in rural areas, which suggests that rural squirrels may be more sensitive to risk and thus they may be more affected by new urban developments, particularly if food is not supplemented. Giving-up densities can even be mapped out in order to identify spots across a landscape with greater risks, which can provide a relatively easy (but indirect) way of understanding spatial variations in the predation costs of foraging (see **Box 6.1** for an example) and, eventually, in how suitable different spots can be for foraging purposes (**Figure 6.2**).

It is important to note that even though giving-up densities can be good measures of the suitability of a landscape, there are many factors that can influence risk perception, such as group size, distance to protective cover, and distance to cover that obstructs the detection of predators. Furthermore, variation in giving-up density does not necessarily reflect variation in the rate of predator attacks or predator behaviors. These factors must be considered when interpreting giving-up density findings. More importantly, giving-up density is a single dimension in the complex process of animal foraging; as such, it cannot be considered a proxy for how animals can digest different types of food, their food preferences, the nutritional value of

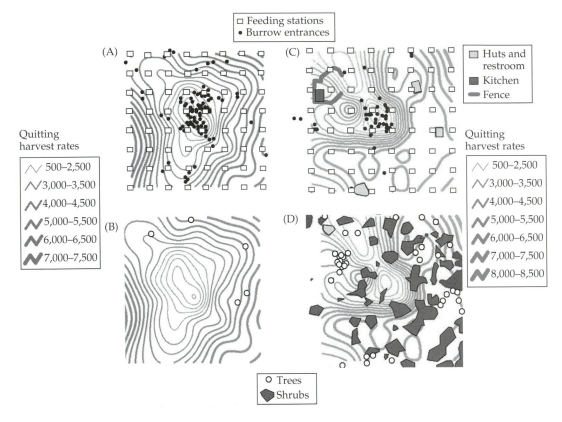

FIGURE 6.2   Landscapes of fear of the (A and B) open and (C and D) shrubby colonies of squirrels. Lines represent areas of similar quitting harvest rates (J/min). Changes in line weight represent variations in the predation costs of foraging. The locations of feeding stations, burrow entrances, and human structures are overlaid in (A) and (C), whereas the locations of shrubs (polygons) and trees (circles) are overlaid in (B) and (D). (After van der Merwe & Brown 2008.)

different food items, the speed with which animals gather food, etc. These other foraging processes may need to be studied independently.

Constraints in Gathering Foraging Information

Animals obtain information about foraging from different sensory modalities. For instance, snakes may track the thermal signatures of their prey, birds may be attracted to a food patch by specific vocalizations from conspecifics, and some raptor species may follow the urine traces of their mammalian prey by using their ultraviolet photoreceptors. This creates an interesting problem for conservation biologists: it is important to understand the

BOX 6.1 Quantifying the Landscape of Fear

Variation across space in the predation costs of foraging creates a landscape of fear (Laundré et al. 2001). Van der Merwe & Brown (2008) developed a useful methodology to map out the landscape of fear of ground squirrels (*Xerus inauris*) that can be extended to other species. One of the most important assumptions underlying this approach is that the metabolic costs and the missed-opportunity costs are constant across the landscape. This is not likely to be the case in a field scenario, because different animals will have different energetic states, and the microhabitat features surrounding the food patches will vary.

Van der Merwe & Brown identified two main steps to create a landscape of fear map. The outcome will be variations in raw giving-up density values, but if we wish the output to be quitting-harvesting rates (expressed in joules per minute) to make it comparable between study areas and organisms, an extra step is required (van der Merwe & Brown 2008).

Step 1: Create feeding stations that will be suitable for the species of interest. Add food and a mixing substrate in amounts that do not vary across feeding stations. Initial applications of this method focused on seed-eating desert rodents. Dried seeds were mixed with sand and placed in a container (aluminum baking pan) to create a homogeneous patch. Giving-up densities work very well for grainivores. The method can also be used for species that eat other foods, but it is essential to be able to thoroughly mix the food with some sort of substrate. This makes animals work to obtain the food, and thus the amount left reflects the trade-off they make between risk and additional efforts to obtain food. There are two things to be aware of when you use giving-up densities to quantify risk. First, you must realize that prebaiting locations is essential. It is not typically possible to simply go out and establish feeding trays; animals may have to learn that food is in particular locations. Second, it is important to ensure that only one species (ideally only one individual) forages on an artificial patch. For nocturnal desert rodents, biologists swept smooth the surface of the sand each evening and then read the tracks each morning. Patches with evidence of other foragers were discarded.

Step 2: Use a map of the study area and create a grid. Keep in mind that the study area should represent the home range of the population of interest, and other populations should not be included in the same grid; otherwise, there can be a sharper variation in the costs of missed opportunities. Once the grid is created, locate the feeding stations. This is a key component, and it depends on the criteria developed by an expert on the species of interest. If feeding stations are too close to each other, there may be little variability between them, so there may be a nonoptimal investment of logistic resources. If the feeding stations are too far from each other, it may not be possible to relate changes in giving-up density to important variation in landscape elements (e.g., slope, distance to burrow, distance to cover). Furthermore, feeding stations must be close to the centers of activity areas. This is particularly relevant for social species, which may have burrows or roosts, so their giving-up densities will be sensitive to slight changes in perceived predation risk. It is

(*continued on next page*)

important to map the landscape elements animals use frequently, to interpret the results appropriately. Feeding stations should be left for long enough that they can be visited several times and can generate variability in the responses of the model species. Van der Merwe & Brown (2008) allowed gray squirrels to feed for two days, removed and counted the seeds left, and calculated giving-up densities. The feeding stations may be placed again in the same spots to get another set of giving-up density measurements. If so, the average of all 2-day giving-up density estimates should be taken per spot, or the design should be treated as a repeated-measures design. All feeding stations in a given area should be active simultaneously to minimize variations in missed-opportunity costs. Giving-up densities should be expressed as grams of food per feeding station. If other species are likely to feed at the stations, the feeding stations may need to be modified to prevent the other species from entering (e.g., by making holes match the size of the species of interest).

Step 3: Build a contour map based on the giving-up densities. Contours will represent lines of equal giving-up density, and variation in the "heights" of the giving-up density contours will illustrate changes in the predation costs of foraging. Different programs are available to build contour maps (e.g., ArcGIS from ESRI—www.esri.com), whose output can be used to estimate the percentages of area that the animals perceive as having high and low levels of risk. Maps of other landscape elements (e.g., trees, bushes, areas with high frequency of recreational activities) can then be overlaid on the contour map to establish whether their spatial location influences perceived risk of predation.

Two of the areas studied by van der Merwe & Brown (2008) nicely illustrate the use of the landscape of fear concept (see Figure 6.2). The shrubby colony was found to have twice the number of feeding stations with high predation cost (>5,000 J/min) in the landscape than the open colony. One implication is that raising the amount of human disturbance may increase the predation costs of foraging much more in the shrubby colony than in the open colony. Feeding stations farther from the burrow increasd the perception of risk. However, risk perception was influenced by distance to trees and shrubs (positive relationship) and grass cover (negative relationship) only in the shrubby colony. This suggests that management measures, such as planting trees to reduce the distance between burrows, would be effective in only one of the colonies.

For species of conservation concern, quantifying the landscape of fear can be essential information that will allow creative management strategies to reduce risk perception, increase foraging intake, and ultimately increase survival and reproductive success. This sort of behavioral study also can be cost-effective if it shows that expensive management strategies (such as planting trees to reduce risk) work only in certain environments.

sensory world of the target species in order to ensure the access to cues that signal food availability (see Chapter 5). This knowledge is essential when dealing with introduced species or with captive individuals in a zoo. For instance, juvenile komodo dragons (*Varanus komodoensis*) are known for their arboreal habits and reliance on both visual and chemical cues to catch

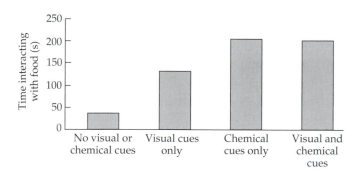

FIGURE 6.3   Time hatchling komodo dragons allocated to interacting with stimulus boxes that contained different combinations of visual and chemical cues. Although komodo dragons used visual cues, chemical cues increased the time spent with food. The study was conducted at the Denver Zoological Gardens. (After Chiszar et al. 2009.)

insects and lizards. However, which cues are more important when it comes to feeding hatchlings reared in captivity? A recent study found that hatchlings in a zoo used visual cues to a limited extent, such as to initiate exploration, but that chemical cues were more important when searching for live prey (**Figure 6.3**; Chiszar et al. 2009).

There are many important questions to answer when dealing with a species of interest. First, what are the different sensory modalities involved in food finding? Second, what is the relative role of each of these sensory modalities (which ones are involved in finding food patches, in finding food within a patch, and in choosing between different types of food)? Third, what are the properties that constrain a given sensory channel during foraging?

These appear to be more basic than applied questions; so does that mean that conservation biologists need to become sensory ecologists as well? Not necessarily. It means that conservation biologists should seek this information in the literature (or collaborate with sensory ecologists) before implementing strategies that seek to attract animals to or repel animals from food patches. *We cannot overemphasize the relevance of understanding the sensory world of the species of interest.* It is the *animal's* perceptions of habitat that are important, not ours, and we should be careful to not make habitat manipulations without a clear knowledge of how a species perceives its "own" habitat.

Let's work on a hypothetical example. The yellow cardinal (*Gubernatrix cristata*) and the black-backed grosbeak (*Pheucticus aureoventris*) are two seedeaters that live in Argentina and are threatened as a result of the pet trade. Let's imagine that a program to bring the populations of these species back from the brink of extinction includes captive breeding followed by reintroduction of individuals in natural and human-dominated landscapes. One way of involving the public is to encourage people to set up bird feeders for these species. This can lead to a citizen science project to monitor the recovery of the species. The idea is simple: give the preferred food to these species, let them forage, and wait until they habituate to humans so that people have them in their backyards rather than as pets in cages.

Now the question becomes: How can we design a bird feeder that is attractive for these species and in which they feel comfortable foraging? We know that birds are highly visual when it comes to their foraging behavior, which addresses many of the aforementioned questions regarding the types of sensory modalities involved. However, within the visual system, there are many properties that can be species-specific and influence the suitability of different feeders. For instance, feeders need to be detectable from far away. And, once in a feeder, the bird must have enough visual coverage to monitor the environment for predators. Finally, the seeds need to be mixed in a pattern that allows them to be recognized as food.

Avian vision is different from human vision (Cuthill 2006). Most birds have eyes that are more laterally placed. This gives the birds both binocular visual overlap and lateral views of the environment. And they have four types of visual pigments (and five types of cone photoreceptors: four single cones and one double cone), whereas humans have three visual pigments. Birds also have oil droplets located in the photoreceptors that filter light before it reaches the visual pigments. The differences in visual pigments, photoreceptors, oil droplets, and ocular media allow birds to perceive colors differently than humans. More importantly, there is a high degree of variability in these visual properties between bird species. Therefore, we must study how each species perceives its surroundings, along with characterizing its habitat.

For instance, let's ask a simple question: What color should the bird feeder be? Unfortunately, we cannot determine which colors birds can really perceive, but we can use models to predict how objects with specific spectral properties will be perceived in relation to a specific visual background, given a set of photoreceptors that have specific peak spectral absorbance. To do so, we need to collaborate with physiologists or use the literature to get estimates of visual pigment sensitivities. If we go the physiological route, we must first examine the retina with a microspectrophotometer and determine the wavelengths with the highest sensitivity in the different types of visual pigments and oil droplets. This information will help us define a multidimensional space of wavelengths the species is probably sensitive to. But simply because a bird can detect a particular *wavelength* very well does not mean that it can detect a colored object *under certain light conditions*. Environments vary in the amount of ambient irradiance. Imagine a forest with light beams shooting between leaves. A feeder will look different in the shade than when lit by a sunbeam. Thus, we must produce a set of candidate feeders of different colors, measure their reflectance as well as the reflectance of the visual backgrounds in the habitats where we want to use them, and finally measure the ambient irradiance in those habitats. Using all of this information (photoreceptor sensitivity, feeder reflectance, background reflectance, ambient irradiance), we can estimate the chromatic contrast of the feeder in the environment. Chromatic contrast is a measure of how well the feeder stands out against its background. Thus, we can make predictions about feeder colors that will increase the chromatic contrast for a target species. We can use a similar approach to the colors of the seed mixtures that have nutritional value for these species.

Another simple question: What is the best shape for a feeder? While animals are in the feeder, their time will be devoted to searching for the preferred seeds, pecking at them, and monitoring the environment for potential predators. Measuring the configuration of the visual fields of the model species will allow us to determine its degree of visual coverage. If the feeder blocks part of the animal's visual field, then the animal may spend more time being vigilant than it would at a natural food patch to compensate for the lack of visual information while pecking and thus reduce the chances of being caught by a predator. In such situations, feeders that are designed to reduce obstruction effects (e.g., thin, long feeders with excellent peripheral visibility) may reduce the perception of risk and make the feeder more likely to be visited and may increase the time the animal spends in the feeder.

The "simple" problem of designing a feeder illustrates the myriad of ways that it is important to account for the sensory modalities animals use when foraging. Remember that we can *exploit* sensory modalities to increase the ease with which individuals find a patch. Interestingly, the same idea can be reversed, and we can exploit a sensory modality to *repel* animals from a given foraging patch, to keep unwanted species from particular areas. For instance, there has been quite a bit of work on use of secondary antiforaging compounds that contain anthraquinone as the active ingredient. Anthraquinone absorbs the UV part of the spectrum and it may be visible to some birds; these birds can develop a learned avoidance (due to latent malaise) to resources treated with products containing this ingredient (**Figure 6.4**; e.g., Ballinger 2001; Werner et al. 2009).

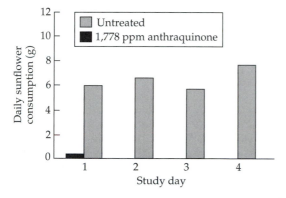

FIGURE 6.4 Consumption of sunflower seeds with and without anthraquinone by red-winged blackbirds (*Agelaius phoeniceus*) in captivity. Anthraquinone can act as an avoidance conditioning agent for birds. (After Werner et al. 2009.)

BOX 6.2 Allee Effects

Many ecologists assume that individuals compete against each other and, thus, as population size increases, there will be more competition, and individual reproductive rate will decline. The normal corollary to this assumption is that at low population densities, individuals will have higher reproductive rates because of less competition.

Allee effects, named after the famous population ecologist Warner Clyde Allee, are a form of negative density dependence (Courchamp et al. 1999, 2008). This means that as a population decreases, individual reproductive and survival rates may decline. For instance, Atlantic cod (*Gadus morhua*) were fished almost to extinction. When fishing was stopped, the population failed to recover; both fertilization and juvenile survival are thought to be reduced at low population densities (reviewed in Courchamp et al. 2008). When Allee effects are present, we can infer that the typical assumption that individuals compete and will do better if there are fewer competitors is incorrect.

At a population level, identifying the situations where Allee effects exist is important in modeling population sustainability. At a behavioral level, identifying the mechanisms underlying Allee effects is an important task in conservation behavior. Doing so requires us to look for cases where animals cooperate or at least benefit from each other's presence. Courchamp et al. (2008) is essential reading on Allee effects.

Other Foraging Information Sources: Social Cues

Information about foraging opportunities may come from cues from the prey itself or from cues from other individuals of the same species. Information produced by conspecifics (or even heterospecifics) is called public information.

In a series of experiments, Templeton & Giraldeau (1995) found that European starlings rely on public information (e.g., other individuals foraging successfully in a patch) to estimate the quality of patches. Using public information can give the individuals advantages in terms of either increasing their intake rates or decreasing the variability in finding food-rich patches. If individuals using public information are more successful in terms of survival or reproduction, one relevant question is: What density is necessary for individuals to take advantage of the enhanced information about patch quality? This is particularly important knowledge for rare social species (e.g., parrots) and those species living at the edges of their distribution ranges or in isolated fragments. Let's keep in mind that the lack of sufficient public information available for individuals can lead to Allee effects (**Box 6.2**).

The local density of individuals in a social species is likely to affect group size (Wirtz & Lörscher 1983; Ostro et al. 2001) simply because there are more individuals available to join foraging groups. However, from a

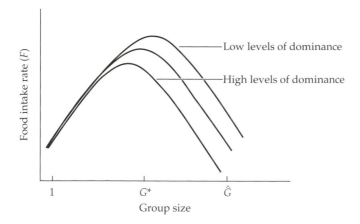

FIGURE 6.5   Optimal (G^*) and stable ($\hat{G}$) group sizes under conditions with variable levels of dominance. At higher levels of dominance, food intake rate decreases for any group size, thus reducing the optimal group size. (After Giraldeau 1988.)

mechanistic perspective, there are trade-offs. If there are too few individuals, there may not be enough to allow individuals to take advantage of public information and enhance food intake; conversely, too many individuals in a group may cause interference and/or quick depletion of resources and thus decrease food intake per individual. Grouping benefits and costs are expected to balance out and give rise to group sizes at which individuals can maximize their fitness returns: the so-called **optimal group size** (Sibly 1983; Clark & Mangel 1986; Giraldeau & Gillis 1984). The curve representing fitness (e.g., measured in terms of food intake or reproductive success per individual) in relation to group size is presented in **Figure 6.5**, where G^* is the optimal group size.

Under certain conditions, a solitary forager joining a group of optimal size G^* can increase its own fitness while the fitness of established members decreases. If this process continues, with one solitary forager joining the group at a time, another point is reached, called the **stable group size** ($\hat{G}$), in which the fitness of new and established members is less or equal to that of solitary foragers (see Figure 6.5). This creates the paradox of individuals paying the costs of group living but without obtaining its benefits (Clark & Mangel 1986; Rannala & Brown 1994).

It has been suggested that the *stable* group size may create an *unstable* equilibrium (Clark & Mangel 1986; Kramer 1985). This is because the addition of new members would reduce the fitness of individuals in the group in relation to the fitness of individuals that decided to forage by themselves. This instability suggests that the stable group size is the maximum group size under a certain set of ecological conditions that would be profitable for an individual. However, the most frequently found group sizes (*realized* or

(A)

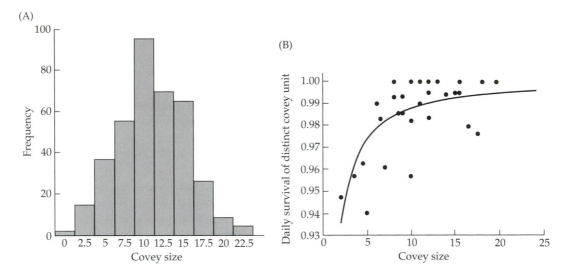

FIGURE 6.6   (A) Frequency distribution of covey sizes and (B) daily survival of different covey sizes of northern bobwhites, *Colinus virginianus*, in Kansas. Covey sizes of 11 individuals were the most frequently observed and showed high survival. (After Williams et al. 2003.)

observed group size) can range between the optimal and stable group sizes (Giraldeau 1988). The stable group size can be equal to or larger than the optimal group size, depending on the shape of the fitness–group size relationship (Giraldeau & Gillis 1984).

Relatively little evidence points to the ecological conditions favoring the occurrence of optimal/stable group sizes. A recent study on northern bobwhites (*Colinus virginianus*), a species that forms semipermanent groups, used a two-pronged approach: first, to establish optimal group sizes in seminatural conditions through changes in predator detection and feeding rates and, second, to assess the variability in group sizes in natural conditions by placing radio transmitters on individuals and estimating individual and group survival, movements, and mass changes (**Figure 6.6**; Williams et al. 2003). The study found an optimal group size of 11 individuals, which was favored because individuals in groups of this size had higher food intake, enhanced predator detection, and higher individual survival. These characteristics ultimately increased group persistence. The bobwhite study is a good example of how to estimate optimal group sizes and the behavioral mechanisms implicated in regulating between-individual interactions.

We can use this information to determine whether the species of interest is below or above the optimal or stable group size in a given area. Then, depending upon where the optimal size lies, we can use different management strategies to shift the size toward the optimum. For instance, if groups are above the stable group size, some individuals may need to be moved to new areas. By contrast, if groups are under the stable group size, habitat can be manipulated to increase the safety of individuals, which will result in enhanced predator detection and higher food intake and survival. Alternatively, reintroducing individuals to a given area where group size is below the optimal (e.g., at the edge of the distribution range) may be considered. Wildfowl managers and those managing deer populations have used these techniques for years.

Other factors may affect the shape of the fitness–group size relationship, such as whether group members limit group membership, the existence of dominance hierarchies, relatedness, intergroup competition, individuals actively recruiting new members into the group, and knowledge about food distribution in the environment (Giraldeau 1988; Higashi & Yamamura 1993; Zemel & Lubin 1995; Beauchamp & Fernández-Juricic 2005). The relevance of these behavioral mechanisms for species of conservation interest requires further research. For instance, greater dominance levels can decrease the optimal group size because dominant individuals may expel certain individuals from the group, which may create a surplus of subdominant individuals. Thus, to ensure the local persistence of such a species, it may be necessary to increase the proportion of suitable habitat in an area and establish new groups from these surplus individuals.

Conclusions

Securing enough food resources is an essential task for animals, but it requires an investment of time and energy while avoiding predators. These costs can affect the payoffs animals obtain from a foraging patch and eventually the chances of using a given patch. When the costs of foraging are higher than the benefits, animals are expected to leave and seek a new foraging patch. We must understand how the costs of foraging vary in space and time so that we can reduce or increase these costs, depending on whether a species needs to be attracted to or repelled from a particular patch. Once an animal is in a patch, it is necessary to learn about how it gathers information that will increase food-finding opportunities. Different species have different constraints on gathering this information, depending on their sensory systems. Understanding these species-specific constraints is relevant to enhancing or decreasing the quality of a food patch and individual prey items, which is a central problem when we manage wildlife. Finally, animals can use other sources of information about the quality of food patches; for example, they can assess the foraging success of conspecifics. Joining groups can bring benefits at small group sizes or costs at high group sizes; the net result is an optimal group size. Learning the mechanisms behind the optimal group size can help conservation biologists determine whether individuals are under- or overexploiting foraging resources and whether extra habitat is necessary.

Further Reading

The edited volume by Stephens et al. (2007) is the best recent entry into the foraging literature. Joel Brown and his collaborators have used GUDs to ask applied questions (e.g., van der Merwe & Brown 2008). Gill et al. (1996) provides a useful discussion of using foraging behavior to study human impacts on wildlife.

7 Understanding Antipredator Behavior for Conservation and Management

Virtually all animals at some point in their lives face the risk of predation. From a management perspective, predators may be responsible for a species decline, the failure of reintroductions, and changes in community composition (e.g., cascade effects). Predation works directly, by eliminating individuals, and indirectly, through the fear it creates in prey, which eventually can influence resource use. Anthropogenic effects on wildlife often work through creating fear. A fundamental knowledge of antipredator behavior may enable us to successfully manage fear and hence manipulate the perceived risk of predation and enhance conservation efforts. For instance, we may develop creative ways to keep individuals away from valuable resources, such as crops, or encourage them to use safe habitats, such as guarded nest boxes. In this chapter we will discuss how knowledge of antipredator behavior is essential for implementing techniques that help reduce predation on vulnerable prey.

Encountering New Predators through Range Shifts and Extinctions

Species ranges change naturally, through range expansions and contractions, and unnaturally, through introductions of predators and prey. The outcome of these shifts may be detrimental for the prey, particularly when naive prey face novel predators.

Species ranges are expected to change as the Earth warms (Davis & Shaw 2001; Huntley et al. 2006), and such changes may cause predators to encounter formerly naive prey. In the past, continental land bridges brought carnivores to North America and facilitated the movement of dingoes (*Canis lupus dingo*) to Australia. These natural or seminatural introductions were associated with extinctions of native fauna (Martin & Klein 1984; Low 1999), although the exact impact of human exploitation, climate, novel predators, novel competitors, and novel pathogens remains an area of active debate (Wroe et al. 2004).

Range changes may be enhanced by human activities. The Channel Island fox (*Urocyon littoralis*), a small endemic fox found on Southern California's Channel Islands, lived for years without native mammalian or avian predators. Compared with their sibling species, the mainland gray fox (*Urocyon cinereoargenteus*), Channel Island foxes were more active during the day and were described as being rather fearless (Crooks & Van Vuren 1995). In the nineteenth century, ranchers introduced domestic livestock, including pigs (*Sus scrofa*), onto some of the Channel Islands. In the 1990s, bald eagles (*Haliaeetus leucocephalus*) went extinct on some of the northern Channel Islands. This extinction was attributed to the failure to rear young as a consequence of chemical dumping in the 1950s and 1960s off the coast of Los Angeles. Bald eagles eat mostly fish, and the fish off Southern California had relatively high levels of DDT and other toxic chemicals that entered the food chain from the toxic wastes dumped in the ocean. About this time, mainland golden eagles (*Aquila chrysaetos*) introduced themselves to the northern Channel Islands by flying 30 km from the mainland. These eagles had a large prey base that consisted of diurnal foxes and feral pigs. It is hypothesized that the pigs sustained and increased the population of golden eagles (there was substantially more pig biomass than fox biomass), which in turn increased predation pressure on the Channel Island fox population, which declined as a result of hyperpredation by the eagles (Roemer et al. 2002). **Hyperpredation** is an ecological process in which species A is being incidentally eaten by predators of a very abundant species B, and the predator's population size is larger than it would be if it were eating only species A. In the case of the Channel Islands, species A was the foxes, and they were being incidentally eaten by golden eagles that were predators of the very abundant species B, the pigs. Managers eliminated the pigs, relocated the golden eagles, and reintroduced bald eagles as part of a comprehensive recovery program that by 2009 had largely succeeded in bringing back the Channel Island fox from the brink of extinction.

Sometimes, the presence of a top predator keeps the population of smaller predators from expanding. The loss of the top predator leads to an increase in the number of midsize predators (also called mesopredators). Since large predators eat large prey and smaller predators eat smaller prey (see Chapter 3), this mesopredator release can be detrimental to smaller prey. An example of **mesopredator release** is found in coastal sage patches in Southern California, where house cats (*Felis domesticus*) and raccoons (*Procyon lotor*) eat songbirds. In patches with coyotes (*Canis latrans*), there are fewer house cats and raccoons (coyotes eat them, cats avoid areas with coyotes, and pet owners keep cats inside) and more songbirds (Crooks & Soulé 1999). However, small coastal sage patches cannot sustain coyotes,

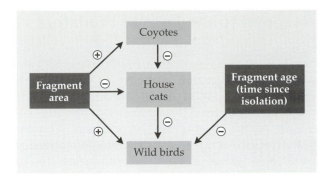

FIGURE 7.1   Mesopredator release in sage patches in coastal Southern California (see text for details). (After Crooks & Soulé 1999.)

and the lack of coyotes increases the abundance of domestic cats that hunt birds (**Figure 7.1**).

Range expansions may also be entirely artificial. For instance, the nineteenth-century "acclimatization societies" in Australia and New Zealand sought to make these British outposts look more like home by introducing British animals and plants (Low 1999). Both predators and prey were introduced by well-meaning, but ecologically naive, citizens. In Australia, introduced predators—red foxes (*Vulpes vulpes*) and cats—are a likely cause of the extinction of at least 18 native mainland mammals since European settlement (Johnson 2006). Cats and foxes were introduced by Europeans in the eighteenth century, and their populations were likely enhanced by European rabbits, which were also introduced in the eighteenth century. Of course, Europeans changed the landscape in many other ways, and the mammalian extinctions could be attributed to a mix of factors (Johnson 2006). However, a large-scale ecological experiment (the Western Shield) conducted in Western Australia employed broadscale fox poisoning, resulting in the later recovery of numerous small mammals (Morris et al. 1998). This experiment demonstrates that introduced predators may be responsible for the small populations of native marsupials.

Red foxes naturally hunt more than they may eat at a sitting. This surplus killing is seen in a variety of carnivores (Short et al. 2002), some of them part of the Australian reintroductions. Short et al. reviewed cases where reintroduced mammals were killed rapidly by one or a few foxes. Reintroduced animals often fall prey to predators, and such predation is one important factor behind reintroduction failure.

Given that species may naturally and unnaturally change their ranges, we should have knowledge about how populations respond to the loss of their predators, how individuals respond to novel predators, how prey species reduce and otherwise manage predation risk, and how such predation risk assessment and management is studied. We discuss these topics below.

How Populations Respond to the Loss of Predators

When species are isolated from their predators, they are expected to lose costly antipredator behavior. For instance, in guppies, bright color spots are sexually selected by female choice; females typically prefer to mate with bright males. However, the level of predation affects both this preference and the trait: guppies living in piscivore-free streams are more brightly colored than those living in piscivore-rich streams. The introduction of predators to a formerly piscivore-free stream leads to a rapid evolutionary response in color; after several generations, only dull-colored males survive (Endler 1995). When predators are removed, there is a rapid evolutionary response, again driven by sexual selection, which quickly leads to a population of brightly colored males. Thus, predation risk is an important factor that puts a check on the elaboration of sexually selected traits.

However, some traits persist for a remarkably long time following isolation (Lahti et al. 2009). North American pronghorn antelope retain a sophisticated set of antipredator behavior that seemingly evolved in response to cursorial predators that are long extinct (Byers 1997). Some populations of California ground squirrels (*Spermophilus beecheyi*) retain rattlesnake recognition abilities despite more than 70,000 years of isolation from rattlesnakes (Coss & Goldthwaite 1995; Coss 1999).

How might species lose their predators, and thus their antipredator behavior? Isolation on islands is a common mechanism. The "tameness" we see in insular species may be in response to isolation from predators or from never having predators to begin with (Curio 1966). Other processes may enhance the loss of antipredator behavior (Blumstein & Daniel 2005). For instance, bringing animals into captivity for captive breeding inevitably leads to a rapid evolutionary response: animals that are unable to tolerate captivity do not breed. If those that tolerate captivity do so because they are generally less reactive to stressful situations, we can envision a rapid and correlated loss of antipredator behavior. Husbandry processes may also isolate animals from the stress generated by a predator, which may eventually lead to a loss of antipredator behavior.

Like most traits, antipredator behavior may have some degree of "hard-wiring" and may require some degree of experience for its proper performance (Blumstein 2002). Thus, isolation from predators, for even a single generation, can lead to the complete loss of experience-dependent antipredator behavior. However, more generations of isolation are required for the evolutionary loss of some more-hard-wired behaviors.

Having Some Predators May Improve Ability to Deal with Novel Predators

Given that not all species lose antipredator behavior following isolation, how may we predict which species might be most vulnerable to novel predators? The multipredator hypothesis (Blumstein 2006b) predicts that when species encounter novel predators, those that already have some predators should fare better than those that have none.

For instance, New Zealand has no native terrestrial mammalian predators and very few, if any, aerial predators. Thus, before human habitation, New Zealand birds had limited exposure to predators. The introduction of terrestrial mammalian predators to New Zealand has decimated native fauna. Contrast this with North American birds: A variety of terrestrial mammalian predators, arboreal predators (mammals and snakes), and aerial predators have selected for a rich suite of antipredator behavior. As species' ranges change in North America, we do not expect the widespread extinction that we saw in New Zealand. How do multiple predators protect a species from novel predators?

The multipredator hypothesis expects that antipredator behavior will be "packaged" or otherwise linked; it should not evolve independently. Consider an ungulate that, when born, reduces predation risk by combining its crypsis with immobility. If these traits evolved independently, individuals with one but not both of them would be at a selective disadvantage. Moreover, to escape its predators when older, that same ungulate must run very quickly. However, in order to be able to run quickly, it must first survive its early life, and individuals that were not cryptic and immobile early in life would be at a selective disadvantage. Also consider that young ungulates face risks from aerial predators (eagles) and terrestrial predators (canids and felids). Individuals that responded appropriately to canids but not eagles would be at a selective disadvantage. Thus, we should generally expect antipredator behaviors to be linked, and species living with multiple predators may have evolved specific antipredator behaviors in response to each of them, but their expression is not predicted to vary independently. Thus, if a single predator is lost, antipredator behavior for predators no longer present should be retained (Blumstein 2006b). However, consider the caveat in **Box 7.1**.

The multipredator hypothesis has been tested with the tammar wallaby (*Macropus eugenii*), a small, macropodid marsupial (Blumstein et al. 2004a). Tammars were studied in a mainland Australian population where they faced the risk of predation from mammalian, avian, and reptilian predators. Tammars from this population had a sophisticated set of antipredator behaviors including group size effects (**Figure 7.2**), by which individual vigilance is reduced when animals aggregate. There are two major explanations for this "safety-in-numbers effect": (1) the individuals could be part of a selfish herd, whereby they would be less likely to be targeted (Vine 1971), or (2) by being with others, they could detect predators earlier (Pulliam 1973). Group size effects are antipredator behavior that should be particularly effective against mammalian predators if by aggregating, animals can detect a predator sooner. Group size effects would be less effective against a sit-and-wait predator such as a snake. Group size effects *might* be effective against raptors through simple dilution effects. Dilution effects are seen

BOX 7.1 A Caveat about the Relative Costs of Maintaining No-Longer-Functional Behavior

Maintaining no-longer-functional behavior will influence the likelihood of linkage and the likelihood of persistence under relaxed selection. Strong selection against linkage should break apart or prevent the formation of linkages. Imagine a linked complex of three different predator-specific antipredator traits—A, B, and C. Assume that the benefits associated with their proper performance are identical; when properly used, they reduce predation risk. However, each will have a cost that emerges in the absence of the predator for which it evolved. For instance, galliformes are often immobile after detecting a raptor (Evans et al. 1993). If there are no raptors but an individual falsely identifies a raptor, it will, by freezing, experience a missed-opportunity cost. Suppressing foraging for a few minutes while responding to a nonexistent predator may not substantially reduce fitness. However, costs may be more extreme than a simple missed-opportunity cost. Many species (e.g., sticklebacks, *Gasterostreus aculeatus*) have metabolically expensive protective body armor in areas with predators (Reimchen 1994). By investing in armor, animals allocate energy that could be otherwise used for reproduction. Thus, there are clear differences in the costs of missed opportunities (most of which are likely to be relatively small) compared with the costs of maintaining armor in a predator-free location (which may directly affect reproductive success). These costs should influence the likelihood of pleiotropy and/or linkage and therefore the likelihood that antipredator adaptations will persist after the removal of one or more predators.

Assume that A and B have identical costs (e.g., missed-opportunity or other time costs) but that C is much more costly (e.g., body armor). The removal of the predator that has selected for behavior C will make the maintenance of behavior C extremely costly. Thus, we might not expect that costly traits will be incorporated into coevolved complexes of antipredator behavior.

More generally, the removal of a predator that selects for a high-cost response will have different effects than the removal of a predator that selects for a low-cost response. Over time, species are exposed to different types of predators, and populations are exposed to different subsets of predators (e.g., Reimchen 1994). We might expect pleiotropy and/or linkage to evolve in situations where the costs of different traits are relatively equal. We might also generally expect behavioral traits to coevolve, whereas more costly, and less plastic, morphological and life history traits might evolve independently.

because as animals aggregate, the risk of predation declines at a rate of $1/N$, where N = the group size. For dilution effects to work, it must also be assumed that a predator will target only one individual and that larger groups are no more attractive to predators than smaller groups.

Tammars from two island populations (one with only snakes and the other with only eagles) retained group size effects, while tammars introduced to New Zealand (where they had no risk of predation) 130 years pre-

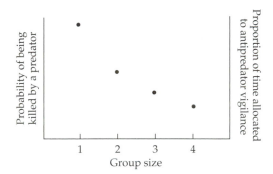

FIGURE 7.2   Group size effects emerge from the decline in predation risk as a function of increasing group size. Through "dilution" effects alone, we expect the probability of predation to decline as a function of increasing group size. Group size effects are seen if time allocated to antipredator vigilance similarly declines as a function of increasing group size.

viously lost group size effects (Blumstein et al. 2004a). Importantly, tammars from Kangaroo Island, which had only wedge-tailed eagles (*Aquila audax*), retained an ability to respond to the sight of mammalian predators (in this case, a novel red fox) despite being isolated from them for about 9,500 years. Predator discrimination abilities in tammars from New Zealand seemed to be declining.

Additional support for the multipredator hypothesis can be found by examining western gray kangaroos (*Macropus fuliginosis*) living in sympatry or allopatry with predators. Group size effects were not found among western gray kangaroos living on Kangaroo Island, but they were found among mainland populations (Blumstein & Daniel 2002). Because of their size, Kangaroo Island western gray kangaroos have had a limited history of predation since the island was isolated. In contrast, mainland animals have always been exposed to predators.

However, while experience with a predator can prepare an animal for dealing with a sufficiently similar novel predator, animals can and will lose the ability to recognize an extinct predator if it possesses a unique cue that the animal uses to recognize it. Stankowich & Coss (2007) found that black-tailed deer (*Odocoileus hemionus*) lacked recognition of a spotted felid resembling a jaguar (*Panthera onca*) 600,000 years after it had disappeared from the area, despite extant puma predation. Interestingly, deer responded strongly to a model tiger even though they had never experienced a vertically striped predator in their history (**Figure 7.3**). It appears that while the vertical tiger stripes did not sufficiently disrupt the general evocative felid body shape in that environment, the jaguar rosettes had lost their salience as a recognition cue for the deer and had become an effective camouflage for the predator.

The multipredator hypothesis also explains the persistence of anti-predator behavior in California ground squirrels (Coss 1999) and pronghorn (*Antilocapra americana*) (Byers 1997). While Byers suggested that the "ghosts

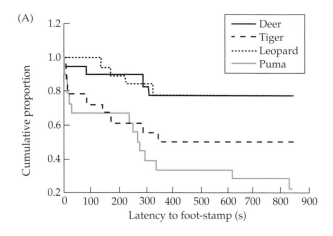

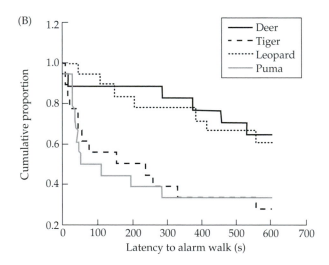

FIGURE 7.3 Colombian black-tailed deer did not have antipredator responses to the sight of deer or leopards but responded fearfully to both tigers and pumas. Illustrated are survival curves plotting the cumulative proportion of the deer responding over time after being exposed to a stimulus. (After Stankowich & Coss 2007.)

of predators past" maintained no-longer-functional antipredator behavior in pronghorn, reconsidering the data in light of the multipredator hypothesis suggested an alternative explanation. Both species (squirrels and pronghorn) still have some predators, thus it would be expected that snake recognition and formerly adaptive responses to now-extinct cursorial predators should persist.

How Do Predators Affect Their Prey?

Predators create fear (Ripple & Beschta 2004; Brown & Kotler 2007), cause anxiety (Berdoy et al. 2000), and may intimidate their prey (Preisser et al. 2005), and these indirect effects can have profound impacts on behavior. This fear is based on a prey's assessment of the risk of predation—formally, the product of the probability of an attack by a predator multiplied by the consequences of such an attack (e.g., Ropeik & Gray 2002). Fear, viewed this way, requires individuals to assess the likelihood of an attack by a potential predator and the consequence (i.e., injury or mortality) of this attack. As fear increases, individuals should attempt to reduce attack probability and/or the consequence of an attack. Thus, predation risk, acting through fear, is a potent selective force that may influence behavioral (habitat selection, resource use, time allocation) as well as morphological adaptations.

Quantifying fear and its consequences in terms of antipredator behavior may provide insights into whether a species is able to respond to its predators adaptively. This knowledge is essential for reducing predation if predators are having deleterious effects on a managed population. For instance, if animals do not recognize a novel predator, predator training may be employed to enhance responsiveness. If animals recognize but do not respond appropriately to a predator, it might be possible to teach the animals to respond appropriately. If such training is not possible, then alternative methods will be required to conserve vulnerable populations.

How Do Prey Reduce Predation Risk?

Individuals may elect to spend less time in areas where predators are common, or they may allocate time differently in habitats that differ according to the probability of encountering a predator. There are several ways to reduce risk while in a risky habitat. Imagine yourself on a trip to an automated teller machine (ATM) in a dangerous neighborhood. You can minimize the time at the ATM by running in, grabbing your money, and running out. Or you can maximize the chance of detecting muggers at a distance by being particularly vigilant while at the ATM. However, by allocating time to vigilance, you are necessarily increasing your time at the ATM in the bad neighborhood. Many species that leave a safe refuge to forage will use either strategy when foraging in a risky habitat; when they decrease vigilance, the total time exposed to predators will decrease, and when they increase vigilance, the total time exposed to predators will increase.

Predation risk is not constant over time. Envision a ground squirrel or a skunk living beneath the path of a raptor migratory route. During the migration season, prey may encounter many thousands more raptors than during the rest of the year. Or consider the daily variation in risk that darkness and moon cycles create. Many species forage more and are less vigilant on dark nights when darkness provides a cloak of safety from visually hunt-ing predators. In an environment where risk varies temporally, how should individuals allocate time to vigilance?

Lima & Bednekoff (1999) developed a risk allocation hypothesis that predicts that (1) animals will be the most vigilant during high-risk situations when such situations are rare, and (2) their vigilance will also be affected by the difference between relatively safe and relatively risky times. Thus, when risks are substantial and rare, it pays to invest a lot in vigilance, and preda-tion risk, particularly temporally variable risk, will impact other activities and structure time budgets. Consider a military squad on a patrol through hostile territory. Since the squad is not on patrol all the time and since patrols are relatively high risk, we expect soldiers to be constantly vigilant on patrol, and we expect them to engage in other activities back in the rela-tive safety of their base. Contrast this with a long, drawn-out battle. At some point, it becomes too costly to perpetually maintain high vigilance, and we expect it to wane. While the risk allocation hypothesis applies generally to any antipredator behavior, behavioral syndromes (discussed in Chapter 9) may prevent individuals from adaptively allocating antipredator behavior as predicted by this hypothesis (Slos & Stoks 2006).

How Do Prey Recognize Predators?

The full range of sensory modalities—visual, tactile, acoustic, and olfacto-ry—are employed to recognize predators. While it may seem initially puz-zling that species should have to learn to recognize their predators, many species do (Griffin et al. 2000). There is a large literature on fish regarding how one experience is sufficient to induce presumably adaptive antipreda-tor behavior (Brown & Laland 2003).

Birds and mammals hone preexisting templates with experience to bet-ter recognize or respond to predators (Cheney & Seyfarth 1990). For instance, tammar wallabies seemingly have a *carnivore* recognition template. In addition to being able to respond to novel carnivores (red foxes; Blumstein et al. 2000), tammars can be trained to enhance their vigilance in response to novel carnivores, but not to novel *herbivores* (goats; Griffin et al. 2002). The existence of such templates may facilitate a number of manage-ment interventions.

Natural development is essential for learning predator recognition, and individuals raised in captivity may face a variety of predator recognition problems. **Priming**, whereby certain experiences make it more likely that predatory stimuli will be evocative, is one mechanism that prepares animals to deal with their predators. For instance, snake-naive monkeys raised with moving food responded fearfully to snakes when later tested, while those raised without moving food did not (Masataka 1993).

Tools to Quantify and Manage Antipredator Behavior

Given this brief overview of some antipredator behavior relevant to conservation biology and wildlife management, how does one actually quantify it? In this section we will introduce some tools that will be useful in studying antipredator behavior. Specifically, we will discuss quantifying habitat selection, studying group size effects, documenting predator recognition abilities, and training prey to respond to their predators.

Studying microhabitat effects on behavior

GIVING-UP DENSITIES   As discussed in Chapter 6 and detailed in Box 6.1, finding giving-up density (GUD), the amount or density of food left in an experimentally provided patch, is a technique used to quantify risk perception. Because foraging decisions are influenced by both the risk of predation while foraging and the lost opportunity to engage in other beneficial activities, animals will leave patches when the rate of return is balanced by the sum of the energetic, predation, and missed-opportunity costs. Thus, individuals might leave a risky patch sooner (and therefore leave more food behind) than they would a safer patch. More generally, as the costs of foraging increase or the values of the harvested resources decline, the harvest rate required to balance costs goes up and more food will be left behind.

Thus, variation in GUD can tell us something about the calculus used by animals. If food "treatment" is fixed and a predation "treatment" is varied (e.g., by putting out feeding trays in different microhabitats, at different distances from cover, or at different times of the year), then the differences in GUD reveal something about the perception of predation risk. Qualitatively this is fine, and while all of this makes intuitive sense, there may be some problems with this approach when one wishes to make precise quantitative predictions (details in Nonacs 2001; Price & Correll 2001).

QUANTIFYING TIME ALLOCATION   For many mutually exclusive situations, animals must trade off one thing for another. For instance, in some species, foraging and antipredator vigilance are mutually exclusive (i.e., animals that are foraging cannot scan the environment for predators). In these cases, animals must allocate time to one or the other activity, and the way they do so reflects the trade-off. Hungry animals may take greater risks, while satiated animals may shun risks. Various factors influence this particular trade-off, but a common theme is that usually it is possible to pay animals to take greater risks (e.g., Abrahams & Dill 1989). Consider military contractors working in a war zone. They are paid many times the salary of someone doing the identical job in a safe location. The amount they must be paid reflects the cost they place on the risk. Similarly, just because an animal forages in a patch does not mean that it is "safe," only that the benefits of foraging outweigh any costs associated with starvation. Thus, one must be very careful before assuming that because animals use a patch, this habitat is "suitable" (see Chapter 5).

Nevertheless, it is possible to identify the relative risks of different microhabitats by quantifying time allocation while animals forage in these habitats. Time allocation can be quantified two ways: by using focal animal samples with continuous recording or by using systematic instantaneous scan samples. Focal animal sampling with continuous recording is conducted by focusing on an individual and continuously recording (for a period of time) all its activities. From this focal record, you can estimate time allocation. By contrast, in scan sampling, observers note, at systematic intervals, the activity of each individual in a group. We prefer the power of focal animal sampling because it provides a richer understanding of vigilance and foraging behavior. In either case, the individual should be the unit of analysis. Thus, it is essential either to have individually identified animals or to ensure that the same individual is not resampled (**Box 7.2**).

For instance, if animals are much more vigilant when foraging in tall grass areas, grass height may influence the perception of risk. Factors to consider include habitat characteristics that influence visibility (e.g., incline, cover, vegetation type), demographic characteristics that influence vulnerability (e.g., pregnant animals may have reduced locomotor abilities), and climatic characteristics that influence the ability to detect predators (e.g., animals are often more vigilant on windy days).

Studying habitat selection

Foraging decisions animals make are influenced by predation risk, and animals trade off the risk of predation with the risk of starvation (Lima & Dill 1990). If human activities create fear in animals, then the decisions animals make about where and how to forage reflect this trade-off (i.e., they will elect to forage in suboptimal locations if by doing so they avoid a perceived risk). Approaches to studying habitat selection are discussed in Chapter 5.

Studying group size effects

If animals rely on safety in numbers, that is, if they have group size effects, then it is important to document this to properly design a translocation or reintroduction effort. Group size effects are not restricted only to highly social species. Much of the literature on group size effects focuses on ephemeral foraging aggregations of birds and mammals that form large groups that may split into smaller units. Thus, it is probably worth determining whether your species of interest has them. Species that have group size effects may profit from being introduced socially, because it may help enhance survival immediately following the translocation (Shier & Owings 2006). Group size effects are studied by regressing group size against time allocated to vigilance and/or foraging. But, how do we quantify aggregation size?

BOX 7.2   Pseudoreplication, Inference Space, and the Study of Animal Behavior

A common assumption of many inferential statistics is that the data points are independent of each other. If we observe the same individual multiple times and treat these observations as independent, we may be pseudoreplicating our data. **Pseudoreplication** occurs when we treat nonindependent data points as though they were independent. In the classic paper on pseudoreplication in animal behavior, Machlis et al. (1985) noted that if you wanted to estimate the running speed of cheetahs, you would not want to run the same cheetah 100 times; rather, you would want to have a sample of 100 different cheetahs, each run once. If you ran the same cheetah 100 times and treated the measurements as independent measurements, you would be grossly inflating your degrees of freedom and would be making an erroneous statistical inference about how fast cheetahs run.

The pseudoreplication issue boils down to an issue about inference space. We define **inference space** as the ability to make generalizations about a sample. If you wish to draw inferences about a population, you must randomly sample that population. If you are more interested in quantifying the abilities of an individual, it is all right to sample an individual repeatedly.

If you do collect multiple observations of an individual and you want to draw inferences about a larger population, you have three choices:

1. You may randomly select a single observation for subsequent analysis.
2. You may average the multiple observations from each individual to create a single value that will be used in subsequent analyses.
3. You may employ repeated-measure statistics, which allow you to draw inferences about a larger population while simultaneously quantifying individual effects (e.g., traditional repeated-measures ANOVA/GLM), and linear mixed effects models (e.g., Steele & Hogg 2003).

Regardless of which method is chosen, it is important to avoid pseudoreplication.

Within an aggregation, individuals influence the behavior of one another. Thus, individuals should be within sight, sound, or smell of their conspecifics. Practically, we quantify group size by counting the number of conspecifics within some distance annulus of a focal subject, in a given area, or within a certain number of body lengths (e.g., Krause & Ruxton 2002). In a series of studies on kangaroos and wallabies, Blumstein & Daniel (2003a,b) found that the area in which the animals were sensitive to conspecifics was larger for larger species compared with smaller species. The largest kangaroos were influenced by conspecifics within 50 meters of a focal subject, while smaller wallabies were influenced only by those conspecifics within 10 meters. To determine the appropriate annulus, count conspecifics within different annuli (e.g., 5 m, 10 m, 15 m) and regress these different group size

definitions against your measures of time allocation. The group size definition that explains the most variation in vigilance and foraging behavior can
be used to quantify group size effects (see Blumstein et al. 2001 for an example). If there are no significant relationships, the species you are studying
may not have a group size effect (or it may not be apparent with the vigilance and foraging parameters measured).

Another approach, using birds, is to conduct seminatural experiments
in which artificial flocks are made up of enclosures and only a single individual is allowed to forage within each enclosure. By keeping the number of
flock members constant, we can manipulate distances between neighbor
birds and assess changes in vigilance and foraging behavior (Fernández-
Juricic et al. 2007a). By comparing the behavior of a focal bird when the
enclosures are next to each other with its behavior when the enclosures are
at increasing distances, we can establish the threshold distance at which
individuals appear to reduce the use of social information and thus may no
longer be in the radius of action of the group. This threshold distance can
later be used as a parameter to measure group sizes in the field (e.g., count
the number of individuals that are within the threshold distance).

When you analyze group size effects, it is important to decide whether
you will average the responses of individuals seen at a given group size or
treat each individual observation as a data point. If there are numerous
important covariates that may vary across repeated observations of an individual (such as distance to protective cover, microhabitat features, or even
the body condition of the focal individual), then it may be better to use each
individual value and include group size as one of several factors in a general linear model or a linear mixed effects model with individual added as a
random effect. However, by doing so, you will reduce the magnitude of any
group size effect. By averaging all observations of individuals foraging while
in a group of a given size (of course, each individual should contribute only
one value to this average; see Box 7.2), you eliminate variation explained by
other variables and therefore make it easier to identify a group size effect. As
with many data analyses, it is often a good idea to analyze it both ways. If
the effect is pronounced enough, you will obtain the same result.

Studying predator recognition

Habitat selection and group size effects are ways that prey can avoid predators and dilute the risk of predation. Many other antipredator behaviors
require that prey first identify a predatory threat. In some cases the lack of
predator recognition may prevent prey from engaging in appropriate
antipredator behavior. How do we determine whether prey can recognize a
predator? We conduct a predator discrimination test. *Discrimination* means
that two (or more) stimuli elicit different responses, while *recognition* implies
higher-level cognitive abilities (i.e., animals form "representations" of different things). Logically, things that are recognized as belonging to different
classes are discriminable.

(A)

(B)

(C)

(D)

50 cm

FIGURE 7.4   Stimuli used to study predator recognition in tammar wallabies. Stimuli included taxadermic mounts of (A) cat, (B) fox, and (C) tammar wallaby, along with a model of (D) the extinct thylacine. (From Blumstein et al. 2000.)

Predator discrimination tests require you to focus on one stimulus modality at a time. Let's take the example of a hypothetical visual discrimination test, which will require a set of visual stimuli and marked animals. Ideally, to avoid pseudoreplication (see Box 7.2), you will need more than one exemplar of each visual stimulus. Practically, this may be difficult, depending upon the stimuli used. For instance, if you are using model predators or taxidermy mounts, it may be prohibitively expensive to make multiple models or to stuff multiple predators. Realize that if you elect to use a single model, your inference space will be limited to the effects of that model on the responses of your study subjects.

What stimuli should you use? You will need at least one treatment stimulus (e.g., a predator model or taxidermy mount), and you will need at least one control stimulus: a nonpredatory stimulus so that you can compare the response to the predatory stimulus with something nonthreatening.

We generally assume that live predators are more evocative than models or taxidermy mounts. However, we use mounts (**Figure 7.4**) because of the control we have over them. It is possible to move the mounts to elicit responses, and some researchers have begun to use remote control models to study both communication and predator/risk discrimination (Patricelli et al. 2006). In addition to having a particular shape, real predators move, smell, and make noises. Why then are we suggesting that you study predator recognition in each modality separately? Because prey may detect predators by each modality alone, it is important to document predator detection abilities in each modality before combining them. This parallels the approach used in the emerging literature on multimodal communication (signals that involve different modalities); to understand them, you must first understand the response within each modality (e.g., Partan & Marler 2005).

Ideally, the nonpredatory stimulus is approximately the same size as the predatory stimuli; if not, the response scored may be a general response to size and not to predatory threat. If predators are novel, you should certainly include a "novelty control"—a stimulus with which the focal species has no experience. By comparing the responses to the predators with the response to the novelty control, you can identify whether novelty itself is threatening. You may wish to have a "presentation control." For instance, if taxidermy mounts are presented on a cart (as they might be for a terrestrial predator), you may want to have a cart control. And, you may wish to have a "blank" control—a period of time during which you go through all the presentation motions (e.g., move around as if you were going to present a cart or a cart with a mount, but do not actually do it) and score behavior.

Predator discrimination experiments work best when they are carried out in captivity. This way you can present the stimuli in a standardized way. Ideally, you will be able to employ a within-subjects design: each individual receives all treatments in either a random order or a systematically counterbalanced order so as to control for order effects (e.g., habituation, sensitization).

What behaviors should you score? You should identify behaviors to score that provide insights into an individual's antipredator abilities. For instance, time allocation to foraging and vigilance may vary as a function of perceived risk. Animals may emit alarm calls (Blumstein 2007b), engage in other specific antipredator behaviors (such as stotting; Caro 1986a,b), or move their heads sideways at different speeds in relation to perceived predation risk (**Figure 7.5**; Jones et al. 2007). Animals may also become immobile. Preliminary observations are essential, and knowing how closely related species respond to threats can also help you develop a set of indicators.

We routinely employ a baseline period: a short time before the stimulus is presented during which we quantify the behavior of the focal subject. This allows us to later subtract out the difference between the response to the treatment and the baseline at the start of the trial. We often attempt to control for initial location and activity by placing a pile of food in a particular location and waiting to begin the baseline period when an animal starts foraging.

When a stimulus is presented, focal observation with all-occurrence sampling (focusing on an individual and recording each behavioral transition) is the best way to quantify the response. We routinely videotape these

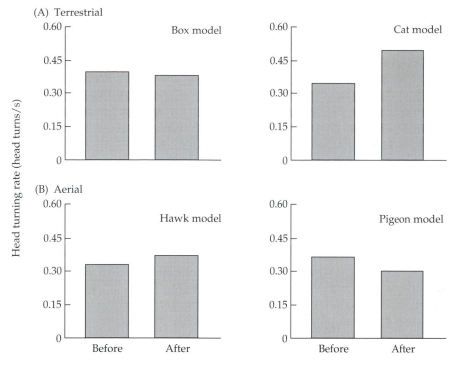

FIGURE 7.5   Rate at which chaffinches (*Fringilla coelebs*) moved their heads sideways before and after being exposed to different types of (A) terrestrial (box and cat models) and (B) aerial (hawk and pigeon models) stimuli. Individuals increased the rate of head movements significantly after being exposed to the cat model. (After Jones et al. 2007.)

experiments and score the videos using an event recorder (e.g., JWatcher— Blumstein & Daniel 2007) at our leisure. It may be possible to dictate responses into a tape recorder and later score these using an event recorder. We do not recommend scoring as the experiment takes place, as there may be some behaviors that could be missed or some problem with the observer (e.g., running out of ink) that prevents data collection.

Depending upon the species and stimulus used, the stimulus may be present only briefly after the baseline period, or it may be present for a long duration. For both mammals and birds, we routinely aim for 1-minute baseline periods and, for visual stimuli, 1-minute visual presentations. We use a 1-minute duration because there is inevitably some movement associated with presenting a visual stimulus. For instance, taxidermy mounts may roll into a viewing window, or a screen may be pulled up. This movement inevitably causes a brief orienting response. Since we are not interested in that orienting response, but rather the response to the stimulus once an individual has had some time to examine it, we leave the visual stimulus in place for a minute. However, the duration of the stimulus will depend on the sensory modality investigated and the specific question to be addressed. Acoustic stimuli often elicit transient responses, and we often only present a few exemplars of them. Olfactory stimuli are usually something that the focal subject must discover, so we leave them until the subject discovers them. It is often difficult to have a baseline period when presenting olfactory stimuli.

Regardless of how you score behavior, you should end up with a series of dependent variables that either reflects time allocation for behaviors that have duration (i.e., they occur as states, e.g., vigilance bouts) or are counts for the behaviors that have no duration (i.e., they occur as events, e.g., alarm calls). Graph each dependent variable against each treatment; some responses will be so rare that further analysis may not be required. The remainder should probably be formally analyzed with general linear models or linear mixed effects models.

To analyze a within-subjects experiment, you should first test for baseline differences as a function of stimuli. If there are significant differences in baseline behavior, then it is desirable to subtract out the baseline time allocation from the response to a given stimulus for each dependent variable. For example, if you are interested in time allocated to foraging, by subtracting the time allocated to foraging during the treatment presentation from the time during baseline presentation, you will create a new dependent variable: difference from baseline in time allocated to foraging. Formally, you will compare this new variable across treatments. Given your set of controls, you will then conduct planned comparisons to compare the treatment with each control (because you have established a set of a priori predictions about how animals should respond to the different treatments), and in some cases, you will conduct planned comparisons to compare certain controls with each other.

In general, you may detect significant differences in the overall response over the period of time the stimulus is present, or you may detect stimulus-specific changes in time allocation over time. Thus, you should consider testing for both "main effects" of a treatment and interactions over time. **Figure 7.6** illustrates this.

Beware of both "floor" and "ceiling" effects! If the stimulus is not sufficiently evocative, there may be no response—a floor effect. By contrast, if the stimuli are too evocative, it may not be possible to identify differences between them because the responses go "through the ceiling." It may be possible to reduce the magnitude of an effect by reducing the risk associated with a stimulus: reduce the size, increase the distance, reduce the volume (for acoustic stimuli), or reduce the concentration (for olfactory stimuli). As already explained, a preliminary study should help establish the existence of floor or ceiling effects.

Training prey to respond to their predators

If prey respond inadequately to predators or if they fail to respond to predatory stimuli, it may be possible to train animals to recognize predators as

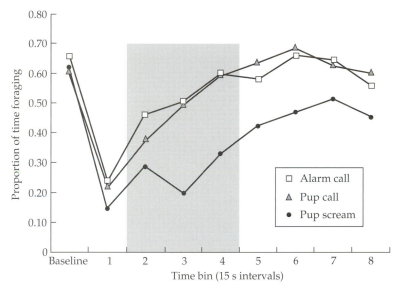

FIGURE 7.6   The proportion of time marmots allocate to foraging before (baseline) and after (in 15-second time intervals) hearing an alarm call from an adult marmot, an alarm call from a marmot pup, or a scream from a marmot pup. Immediately upon hearing any broadcast stimuli, marmots reduced foraging. The type of the stimulus heard influenced the time it took to recover to baseline foraging. After hearing adult or pup alarm calls, subjects resumed baseline behavior within about a minute, while it took longer to recover after hearing a pup's scream. (From Blumstein et al. 2008.)

threats and to respond more aversively to them (Griffin et al. 2000). When this is done before a release, it is called **prerelease training**. Prerelease training may enhance survival upon subsequent release (Shier & Owings 2007a,b), particularly if it allows animals to survive their first encounter with a predator and to learn from that experience.

It is a common misperception that predator training can occur without previous studies of predator discrimination. It is essential to have a comprehensive understanding of predator discrimination abilities before attempting to train prey to respond to predators, for two reasons. First, it is costly, and not without risk, to train animals to respond to predators. One cost of training is habituation—creating an unintended decreased response to the predatory stimuli. Second, you must know how animals respond to stimuli in the first place before you can enhance a response. For these reasons, prerelease training logically follows studies of predator discrimination.

Prerelease predator training boils down to conducting a learning experiment. The psychological literature clearly defines what is required to demonstrate learning (e.g., Shettleworth 1998): learning is seen when a posttest response exceeds a pretest response. However, because simple maturation or other experiences may be responsible for the difference between

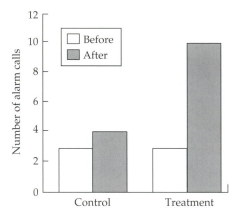

FIGURE 7.7   Analysis of a hypothetical after–before, control–treatment experiment to study learning. In this case, the dependent variable (e.g., number of alarm calls given after being exposed to a stimulus) increased over time for both control and treatment. However, the dependent variable increased more after treatment (10 – 3 = 7) than it did with the control (4 – 3 = 1), illustrating that an individual's behavior was changed by the treatment.

time 2 and time 1, it is essential that you compare this difference between a treatment and a control (**Figure 7.7**).

Treatments in learning experiments are instances in which the trained stimulus (e.g., a taxidermy mount of a fox) is paired with the presentation of something that normally elicits the desired response. Griffin et al. (2001) trained tammar wallabies, to enhance their response to foxes, by presenting tammars with a fox and then chasing them with a net. They did this on four separate occasions. *Controls* in learning experiments are situations in which the trained stimulus (e.g., a taxidermy mount of a fox) is explicitly not paired with something that normally elicits the desired response. In the case of tammars, Griffin et al. (2001) presented the fox at random times throughout the day and chased the tammars at other random times throughout the day.

Comparing the pre- and post-training behaviors of the control group with those of the treatment group will demonstrate whether the training procedure worked. In the case of tammars, not only did they learn to increase their response to the sight of the taxidermy mount of a fox after only four training trials (Griffin et al. 2001), but trained subjects were also able to transmit this response socially to nontrained conspecifics (Griffin & Evans 2003).

While species are primed to learn about certain biologically relevant stimuli, they are unable to learn about all stimuli. In the case of the tammars, they could be taught to become fearful of a novel predator (a fox) but not a novel nonpredator (a goat). This illustration of the **Garcia effect**—

whereby animals are prepared to learn about biologically relevant stimuli (Shettleworth 1998)—tells us that training should work best when a species has had some evolutionary experience with predators. Prerelease training will also likely work when the predatory stimulus to be trained resembles a previously important predator (Griffin et al. 2000).

Animals learn best when an obvious stimulus is associated with an obvious consequence. Thus, it should be straightforward to train prey to respond to the sounds and sights of their predators. This is because the presentation of both acoustic and visual stimuli can be tightly coupled with the presentation of the stimulus that normally elicits an aversive response. It may be much more difficult to pair the presentation of an olfactory stimulus with an aversive response, because it is not known when an individual first can detect the olfactory stimulus. If novel olfactory stimuli elicit investigation, a good first step would be to pair initial investigation (e.g., sniffing) with the alarming stimulus. It may be very difficult to teach animals to be fearful of a particular location (e.g., obstructive cover created by a shrub), because there may be many stimuli associated with cover.

Conclusions

Animals have a variety of ways that they use to reduce the risk of predation. This antipredator behavior has a variety of implications for wildlife management. First, the introduction of a novel predator should be less detrimental when (1) a species already has a similar predator and (2) a species has a rich evolutionary history of exposure to predators. Second, captive-breeding programs that eliminate all risks from prey may be responsible for the rapid evolutionary loss of antipredator behavior. Third, it is essential to understand predator recognition abilities prior to instigating a prerelease training program. Fourth, if predator training is employed to restore antipredator behavior, it should be easier to train species to respond to predators if they have had a history of exposure to predators than if they have not. In addition to these points, a fundamental knowledge of the cues animals use to assess predation risk can be exploited to keep animals off roads and runways, crops and industrial sites, and urban areas. Thus, knowledge of the mechanisms of antipredator behavior allows us to manage when and how animals use space. We will expand upon this in Chapter 12 when we discuss buffer areas.

Further Reading

Caro (2005) is the go-to book on antipredator behavior. Martin & Bateson (2007) provides a succinct summary of quantifying behavior. Blumstein & Daniel (2007) provides a step-by-step guide to quantifying behavior using a freely available computer program—JWatcher (www.jwatcher.ucla.edu).

8 Acoustic Communication and Conservation

Effective communication is essential for many things that animals do. Communication is the glue that cements animal bonds—imagine the soft whispers of courtship or the contact calls between mothers and their dependent young. Communication is often a mechanism through which mate choice decisions are made—think of a female bird listening to and comparing the songs of two suitors before electing to nest with one of them. Communication is a tool used in territory defense—imagine a gibbon troop hooting in their dawn chorus and the forest reverberating with the replies of other troops. These signals all evolved to function well in particular environments.

However, anthropogenic changes in the habitat modify the acoustic environment and therefore may make it difficult for animals to effectively communicate. Such changes may result from modifying the background noise levels (e.g., in cities or along roads), through direct modification of the habitat (e.g., through land clearing or other habitat transformations), or through changes in species composition (e.g., by adding new species). Noise can acoustically mask vocalizations or other animal sounds and thus prevent the message from getting across, or it can disturb animals or even lead to either temporary or more-permanent hearing loss in a particular frequency range (Nowacek et al. 2007).

Species have distinctive vocalizations. Within a species, vocalizations can be distinctive within populations and even within individuals. We may capitalize on this distinctiveness to acoustically census individuals of threatened or rare species. In this chapter, we will discuss both conservation problems and other anthropogenic habitat modifications brought on by noise, and how to take advantage of individuality to survey individuals or species.

A Brief Introduction to Habitat Acoustics

All signals must be transmitted through the environment from a sender to a receiver, during which time a number of physical processes will inevitably modify the structure of the signal through two main processes: **attenuation**, or loss in amplitude, and **degradation**, or loss of fidelity.

Sounds become softer as they are transmitted from a signaler to a receiver. Sound is measured on the logarithmic scale of decibels (dB), relative to a standard value. A jet aircraft engine produces a painful 120 dB sound pressure level (SPL), relative to the threshold of human hearing, 20 micropascals (μPa), while the active sonar the US Navy is developing produces an extraordinary 223 dB, relative to 1 pascal (Pa) in the 2.6 to 3.3 kilohertz (kHz) range and 235 dB in the 6.8 to 8.2 kHz range (Nowacek et al. 2007). It is difficult to make precise comparisons between sound intensities measured over land and in the water with respect to potential damage that might be caused by loud sounds, for at least two reasons. First, sound levels in the water are measured on a different scale because sound transmits differently in the water than over land. Second, each species is sensitive to its own set of frequencies, and damage caused by sound depends on how it is perceived. For instance, because of the logarithmic scale on which sound is measured, US Navy sonar is many times louder than the aircraft engine and is likely to cause behavioral changes and/or temporary or permanent hearing damage. Marine construction and drilling also produce exceptionally loud noises (Nowacek et al. 2007). Natural gas and oil exploration in terrestrial environments creates very loud sounds that can occur without a break for long periods of time. With anthropogenic noises like these, we should generally expect that the behavior of animals exposed to these sounds will be modified, either temporarily or more permanently (Nowacek et al. 2007).

Imagine a bird singing from a tree. For each doubling of distance from the bird beak, there is the inevitable loss of 6 dB of amplitude. If the birdsong is 100 dB at 1 meter from the beak, the song amplitude is no more than 94 dB at 2 meters. This 6 dB attenuation in amplitude is entirely due to **spherical spreading**—the process through which sound waves diffuse through space. Measured at 4 meters from the source, the amplitude should be no greater than 88 dB. As distance from a source increases linearly, the area that the sound covers increases at a much greater rate. Attenuation greater than 6 dB per doubling distance is referred to as **excess attenuation**. Temperature and humidity influence how a sound is transmitted, as do habitat features and vegetation by their refractive effects.

Sounds also lose their fidelity—they are transformed—as they are transmitted through space. Visualize the heat waves reflected off a parking lot surface on a hot summer day. A sound that was transmitted across the parking lot would sound very different than it did at the source. Such **amplitude fluctuations** (changes in amplitude) and **reverberations** (echoes) degrade the sound as it is transmitted from its source. Vegetation will absorb sound, while rocks and other hard surfaces will reflect it.

Because animals have evolved to live in certain habitat types, their vocalizations have similarly evolved to be best transmitted in these habitats. Formally, this is called the **acoustic adaptation hypothesis**. Alone, acoustic adaptation has a relatively small effect on the structure of vocalizations (Daniel & Blumstein 1998; Boncoraglio & Saino 2007). Major differences in habitat structure, however, can generate sufficient selection to modify the structure of vocalizations. For instance, birds that live in the dense forest have more tonal calls that are sung at a slower rate than the calls of grassland birds, which are often "buzzier" and sung at a faster pace. This is because a dense forest will degrade fast and buzzy calls (there will be a lot of reverberation from the sound reflection off the trees), and thus it will select against them.

How Anthropogenic Changes May Change Animal Signals

Habitat structure influences attenuation and therefore excess attenuation. As humans modify the structure of the habitat, through forest clearing, house construction, and road building, the habitat through which a signal passes from a sender to a receiver changes. Vocal communication requires receivers to be able to respond to senders, and this may create a problem because if receivers are expecting different signals, the modified vocalizations may not be perceptually salient.

Some evidence for this comes from the great tit, a European passerine, which sings at higher frequencies where there is enhanced lower-frequency anthropogenic noise (Slabbekoorn & Peet 2003; Slabbekoorn & den Boer-Visser 2006; **Figure 8.1**). Additional evidence comes from the study of song sparrows (*Melospiza melodia*) by Wood & Yezerinac (2006) and the study by Fernández-Juricic et al. (2005b) of house finches (*Carpodacus mexicanus*), which adjust their vocalizations so that they are effectively transmitted through urban noise. Songs are not the only vocalizations that may need to be adjusted; we should expect that vocalizations and acoustic signals used in other contexts may be similarly affected (Warren et al. 2006). For instance, orcas (*Orcinus orca*) increase their vocalization rates in response to the anthropogenic noise created by whale-watching boats (Foote et al. 2004).

Habitat modification also may influence species composition (the identity of the species present in a habitat). Species do not live, or communicate, in isolation from other species. Imagine the constant din created by many species of chorusing frogs in a pond in the spring. The **sound environment hypothesis** predicts that acoustic features that are biologically important to a species should be modified so as to avoid being masked by other biological sounds (Marler 1960). Ultimately, the sound environment hypothesis predicts that changes in species composition should lead to changes in the structure of vocalizations.

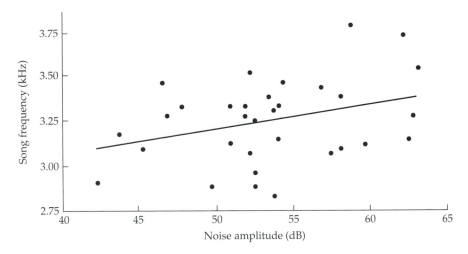

FIGURE 8.1   The effect of urban noise on the fundamental frequency of the songs of great tits. As noise amplitude increased, the fundamental frequency of song increased so the tits could sing over the urban noise. (After Slabbekoorn & Peet 2003.)

There are two ways that changing the acoustic selective regime by introducing novel species or sounds may be detrimental. First, changes in what signalers are producing may be irrelevant to receivers if they did not coevolve with them; they may literally fall on deaf ears. An example is a species that has evolved to communicate to its predators to discourage pursuit (e.g., rodent alarm calls, which may warn conspecifics about the presence of a predator, initially evolved as a signal to discourage predators' pursuit; Shelley & Blumstein 2005). If there is a novel predator that does not respond to the prey's signals, individuals trying to communicate to the novel predator may be in trouble. Second, individuals may be unable to modify the structure of their vocalizations in a way that ensures their signal is transmitted through the novel sounds that are produced by newly introduced species or directly by humans. Such modifications can happen through an experience-based mechanism (e.g., learning) or over evolutionary time (Patricelli & Blickley 2006), but we should not expect all changes to be immediate or equally possible.

Thus, it is essential to understand the potential of species to modify their signals. Some modifications may be easier than others. For instance, when you are in a crowded and loud area, you naturally talk louder. Such **Lombard effects** (Lombard 1911) are reported in a variety of taxa (Pytte et al. 2003; Brumm 2004; Brumm et al. 2004; Scheifele et al. 2005; Egnor & Hauser 2006; **Figure 8.2**). However, we do not know the degree to which species are able to overcome masking by other means, such as frequency shifting, microhabitat selection, or changes in the time when they are active. Answers to such questions will be of great interest to conservation biologists and wildlife managers.

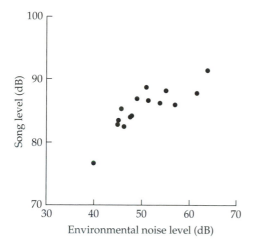

FIGURE 8.2 Example of the Lombard effect in the song of male nightingales (*Luscinia megarhynchos*). As the background environmental noise increased, nightingales sang louder songs. (After Brumm 2004.)

Quantifying the Effect of the Habitat on Biological Sounds

Determining whether a species adjusts its vocalizations as a function of background noise is, theoretically, straightforward: record animals vocalizing in different habitats, record background noise, and see if there are associations between the two.

Recording and digitizing sounds for analysis

The objective of all bioacoustic recording is to record sounds as faithfully as possible; recording sounds is best done with high-quality acoustic equipment that is expensive but not exorbitant. Analog tape recorders (either cassette or reel-to-reel recorders) transduce a sound onto magnetic tape. While the highest quality ones are outstanding (e.g., the Swiss-made reel-to-reel Nagra recorder is still used in the film industry because of its high quality and ability to capture a broad dynamic range), analog tape recorders record unintended sounds (e.g., hums and "tape hiss" created by the motors), and unless the motor runs perfectly, it is likely that the recordings will have some frequency fluctuations because the speed is not constant. Additionally, analog recording creates another step: you must later digitize the recordings so that you can quantitatively describe them.

The process of digitizing a sound takes an analog sound and determines the frequencies and amplitudes present in incremental time slices. There are two parameters of interest when digitizing a sound: the frequency at which you sample sounds (in Hz or kHz) and the number of bits of information used to encode amplitude variation—the **dynamic range**.

Compact disc-quality sound is typically sampled at 44 kHz. This means that the frequency at each slice through time is determined 44,000 times each second. Higher-frequency sounds require higher sampling rates. The general rule is to sample at least twice the maximum frequency that you wish to quantify. Sampling at slower rates than this "Nyquist frequency" creates a problem: there will be aliasing—the creation of spurious frequencies in your digitized recordings. Unless you are studying a species with ultrasound, a 44 kHz sampling rate is sufficient.

Compact disc-quality sound is typically 16-bit sound, which means that each amplitude in a sample is quantified by one of 65,536 possible amplitude values (2^{16}). By contrast, 8-bit sounds are described by 256 possible amplitude values (2^8). Thus, the higher the bit rate, the greater the dynamic range. For most studies, 16-bit sound recordings are sufficient.

Digital recorders are preferable to analog recorders because digitizing is done as the sound is being recorded and because there are no recording artifacts. Digital audio tape recorder (DAT) machines transduce analog sounds into digital sounds. They are sensitive units that do not do well in high-humidity environments or where there is a lot of dust. The tape heads are like video camera heads, and they spin at a high speed and are intolerant of perturbations. Moreover, because sounds are stored on magnetic tape, the sounds may have to be redigitized for analysis.

These days, it is best to use recorders that automatically digitize sounds and to save them on a hard disk or compact flash card. Such direct-to-disk recorders can be obtained for less than $1,000, and once digitized, files are easily transferred to a computer for analysis. MP3 recorders and sound files saved in the MP3 format should generally be avoided, since the compression algorithms lose information that is not meaningful to us. However, that lost information may be meaningful to a nonhuman.

High-quality microphones, particularly directional "shotgun" microphones, may set you back another $1,000. Microphone quality is essential because high-quality microphones do not modify, degrade, or filter the sounds you record. Nondirectional microphones should be used to record background sounds, but more-directional microphones may be essential for recording vocalizing animals from afar.

When purchasing a microphone, you should pay attention to the frequency range and the "flatness" of the frequency response. A microphone with a flat frequency response is equally sensitive across a range of frequencies. All microphones "roll off," that is, they become less sensitive at higher frequencies. Practically, this means that the sound must be louder in order to be properly detected. Ideally, your microphone should record the frequencies of the animals that you study without excessive degradation and with equal sensitivity. Thus, your microphone should ideally have a flat frequency response across the range of your species' vocalizations.

Quantifying the structure of sounds

Once vocalizations and background noise are recorded, they must be analyzed. Many analyses focus on determining the frequencies present and their relative amplitudes. Imagine a 2.5 kHz tonal birdsong (e.g., from a varied thrush, *Ixoreus naevius*; **Figure 8.3**). If the species lived in an environment with other 2.5 kHz sounds, individuals singing 2.5 kHz songs might be at a disadvantage. Thus, there should be selection on individuals to sing above 2.5 kHz so as to avoid acoustic masking. In this case, you would want to quantify the frequencies in the birdsong and determine whether ambient or anthropogenic noise was present around 2 kHz.

The Cornell Lab of Ornithology has developed two bioacoustic programs that can be used to quantify sounds. Canary (freely available) runs on OS 9 Macintosh computers and Raven (the full version is sold, but a trial version is freely distributed), written in Java, runs on more modern Macintosh and Windows systems. Both programs are suitable for the quantitative description of animal and environmental sounds.

Sound can be visualized as a **waveform**: a plot of amplitude by time (see Figure 8.3B). However, many bioacoustic questions require the analysis of frequencies. Sound analysis was revolutionized when it became possible to visualize frequency characteristics. Early spectrographs used a series of precisely tuned filters to create a **spectrogram**, or sonogram, an image that plotted time on the *x*-axis, frequency on the *y*-axis, and relative amplitude as a gray-scale *z*-axis (see Figure 8.3A). However, once a sound is digitized, there are mathematical algorithms (Fourier transformations) that can extract the frequencies present and graphically display them. Modern microcomputers

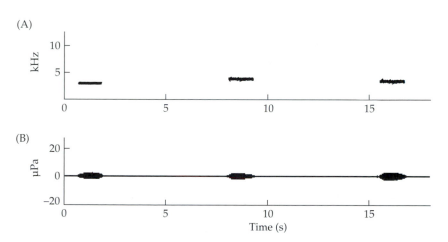

FIGURE 8.3 A spectrogram and waveform of the song of a varied thrush (forest). Note that the (A) spectrogram plots frequency over time and the (B) waveform plots relative amplitude over time.

use a "fast Fourier transform" (FFT) to extract frequency characteristics and draw spectrograms.

There are several parameters you should select when creating a spectrogram. It is essential to understand that when calculating an FFT, you create an unavoidable trade-off: you cannot simultaneously estimate frequency precisely and estimate time precisely. If you wish to estimate frequency precisely, you must calculate what is known as a narrow bandwidth spectrogram. Bioacoustic programs allow you to modify your "filter bandwidth." However, by doing so, you will make your spectrogram temporally inaccurate. By contrast, if you wish your spectrogram to have excellent temporal resolution, you will have to sacrifice frequency resolution. Ultimately, you must experiment with the sounds you are studying and realize that you may wish to redraw spectrograms several times with different parameters to make different measurements.

In addition to a spectrogram, it is often of interest to know which frequencies are present in one or more slices through time. A **power spectrum** is a representation of sound that plots frequency on the x-axis and amplitude on the y-axis. You can think of a power spectrum as a slice through a spectrogram, because if you line up a series of power spectra along their temporal axis, you will create a spectrogram. You will have to make the same decisions about frequency resolution you made when calculating a spectrogram.

With a waveform, a spectrogram, and power spectra, you can quantitatively describe sounds. Temporal measurements should be made off the waveform because of the unavoidable temporal error in the spectrograms and spectra. If your recordings contain background noise, which is usually the case in nature, and because the higher-amplitude parts of vocalizations may be more important for longer-distance communication, it may be more appropriate to measure only the loudest parts of the vocalizations. In some bioacoustic programs (e.g., Canary), this is easily done by calibrating the spectrogram so that you are only measuring the top 40 dB of sound. Frequency measurements may be made on the spectrogram and/or the spectra, but spectra are routinely used to calculate the dominant or peak frequency (the frequency with the most amplitude). Maximum and minimum frequencies present, and thus, bandwidth can be measured off the spectrogram (**Box 8.1**).

Determining whether species can modify their vocalizations to avoid acoustic masking

By systematically recording vocalizations and ambient noise at a variety of locations along a gradient of ambient noise, you will be able to determine whether individuals are able to communicate around background noise. For instance, the green hylia (*Hylia prasina*), an African rain forest bird, must communicate through a variety of insect songs. The hylia is a widespread species, and insect species vary by location. By recording insects and the

BOX 8.1 A Set of Acoustic Measurements Used to Quantify the Structure of Marmot Alarm Calls

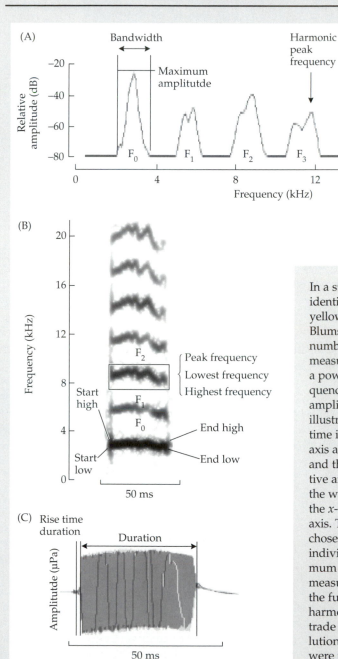

(A)

Bandwidth

Harmonic peak frequency

Number of peaks

Maximum amplitutde

Relative amplitude (dB)

-20

-40

-60

-80

F_0 F_1 F_2 F_3 F_4

0   4   8   12   16   20

Frequency (kHz)

(B)

Frequency (kHz)

20

16

12

8

4

0

F_2

Peak frequency
Lowest frequency
Highest frequency

Start high

F_1

F_0

End high

Start low

End low

50 ms

(C) Rise time duration

Duration

Amplitutde (µPa)

50 ms

In a study of the information about identity potentially contained in yellow-bellied marmot alarm calls, Blumstein & Munos (2005) made a number of time and frequency measurements. Figure A illustrates a power spectrum that plots frequency on the x-axis by relative amplitude on the y-axis. Figure B illustrates a spectrogram that plots time in milliseconds (ms) on the x-axis and frequency on the y-axis, and the gray scale represents relative amplitude. Figure C illustrates the waveform that plots time on the x-axis and amplitude on the y-axis. These measurements were chosen because they varied among individuals. Bandwidth (the maximum to minimum frequency in a measurement) was calculated for the fundamental and subsequent harmonics. Because the FFT must trade off time and frequency resolution, temporal measurements were made on the waveform.

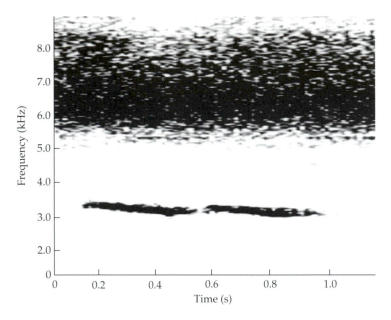

FIGURE 8.4   Green hylia songs are modified to avoid background insect songs. Here we see a hylia song below the broad band of insect noise. (After Kirschel et al. 2009a.)

hylia at a variety of locations, Kirschel et al. (2009a) found that hylia songs varied as a function of the dominant band of insect noise (**Figure 8.4**).

When actually recording noise, pay particular attention to recording the background noise at the same time and location where the animals of interest are vocalizing. This can be done by measuring the noise before or after a species' vocalization from the same recordings used to study the vocalizations themselves. Or it can be done more systematically by measuring the ambient sound at fixed time intervals.

Quantifying hearing abilities

A fundamental understanding of the effects of noise on animals requires knowledge of a species' hearing abilities. Sounds above or below a species' hearing range may not be harmful, in the sense that they will not interfere with communication.

We generally expect a species to be most sensitive to the frequencies it produces. Tuning curves (or **audiograms**) are graphs of hearing thresholds that plot the threshold hearing frequency on the x-axis and the amplitude (in dB) on the y-axis (**Figure 8.5**). The most sensitive frequencies have the lowest hearing threshold (i.e., animals can detect relatively quiet sounds). There are two ways to calculate them: with an operant box or by directly

(A)

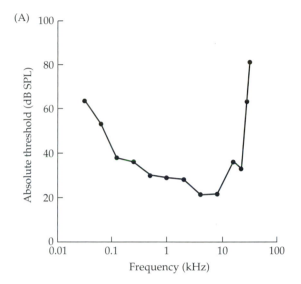

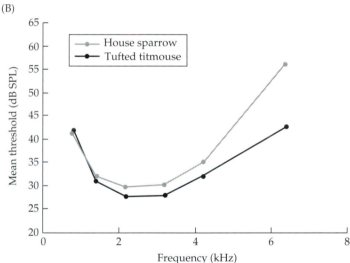

FIGURE 8.5   Audiograms plot the minimum detectable amplitude (in dB SPL) as a function of frequency. Audiograms are species-specific. Illustrated are audiograms from (A) woodchuck (*Marmota monax*) (data from Heffner et al. 2001) and (B) house sparrow (*Passer domesticus*) and tufted titmouse (*Baeolophus bicolor*). Titmice have high-frequency alarm calls that appear to be driving better sensitivity at high frequencies. (After Henry & Lucas 2008.)

recording neural activity on the auditory nerve. Calculating a tuning curve for a species of interest is probably beyond the scope of a wildlife biologist, but you can always collaborate with an auditory physiologist. Having animals in captivity is essential, and studying hearing abilities need not be harmful. Importantly, tuning curves for a number of species have been summarized in one location (Fay 1988). Selection will act on conspecifics to communicate within these most sensitive frequencies. Interestingly, different species may have different sensitivities, which can be a function of variations in the frequency of calls (Henry & Lucas 2008; see Figure 8.5B).

Applications

Anthropogenic noise can be particularly harmful at times when a species must communicate (some birds and many primates sing in the morning, and frogs may call at dusk or all night long) or when it affects the specific frequencies that individuals use to communicate. Extremely high-frequency sounds or extremely low-frequency sounds are omnipresent and may not interfere substantially with communication. If the sounds created by machinery are not within the bandwidth of the species of interest, then that noise may not be detrimental to those species. Or if anthropogenic noise is created at times of the day when the species of interest do not vocalize, then it may not interfere with their communication.

However, this does not mean that the noise is not disturbing in other ways. For instance, in apparent response to the extraordinarily loud active US Navy sonar, dolphins and whales swim away from the sound source, rapidly change their depth (which results in what in humans would be called the bends), and swim ashore (creating what have been called mass strandings) (Jasny et al. 2005). And it is possible that the noise may interfere with animals in other ways as well. More bats are killed by wind turbines at wind farms than might be expected, given their superb flight abilities. Because they breed relatively slowly and have very small litters, this additional mortality may be significant (NRC 2007). Baerwald et al. (2008) found that 90% of bats' deaths near wind energy facilities were the result of barotrauma caused by rapid air-pressure reduction near moving turbine blades, whereas a small proportion of deaths were caused by the collision of the animals with the turbine blades themselves. It is also possible that bats are susceptible to the high-frequency noise created by wind turbines that may either attract bats to the turbines or disorient hunting bats, which may be foraging on insects that might be attracted to the warm turbine blades (NRC 2007). We believe that one should adopt the **precautionary principle** (i.e., it is better to be safe than sorry and the onus is on those who want action to demonstrate that action will not be harmful) when dealing with anthropogenic noise. For example, we should assume that bats may be affected by turbine noise, and we should not use scientific uncertainty as a reason to postpone action to develop ways to prevent bats from hitting turbines. Knowing something about how and when animals communicate may suggest ways to reduce disturbance. Wildlife biologists are optimally positioned to conduct these sorts of studies.

Capitalizing on Species-Specific and Individually Specific Vocalizations

Rare species, by their very nature, are often difficult to detect. Trapping is used to estimate species diversity, and **mark-recapture techniques** are routinely used to estimate population sizes. Assume you make two visits to a location and trap or otherwise detect individuals in a systematic way. If you assume no immigration, emigration, or mortality between the two sampling days, and if you assume that all individuals are equally detectable or able to be trapped, then you can use these two samples to estimate the population. The Lincoln–Petersen method (Krebs 2001) estimates the population size as

$$N = \frac{n_1 n_2}{m}$$

where N = your estimate of the total population size, n_1 = the total number of individuals captured or detected on the first visit, n_2 = the total number of individuals captured or detected on the second visit, and m = the number of individuals captured or detected on the first visit that are also captured or detected on the second visit. The logic is that the ratio of the total number of individuals captured or detected on the second visit to your estimate of the total population size (n_2/N) is the same as the ratio of the number of individuals captured or detected on both visits to the total number of individuals captured or detected on the first visit (m/n_1).

Because there is strong selection for vocally communicating species to sound different so as to avoid directing their messages to the wrong species, calls and songs are routinely used by biologists to detect the presence of a given species during bird, mammal, frog, and insect acoustic surveys. In addition to species-specific variation, many species have individually distinctive vocalizations. This variation arises from morphological variation in sound-producing organs as well as postproduction filtering that modifies the structure of vocalizations (Fitch & Hauser 2002). Selection may act on social species to produce even more-distinctive vocalizations (Pollard & Blumstein 2010). Regardless of the variation's proximate basis or ultimate benefit, in vocally communicating species, it is possible to identify individuals by their vocalizations, and thus it may be possible to census the individuals of a given species using recording and rerecording protocols. To do so, we should be able to (1) quantify individuality and (2) use this knowledge to identify individuals vocally.

Individually distinctive vocalizations in social species

Sociality is characterized by relationships between individuals (Wey et al. 2008). Thus, individuals must be able to discriminate and identify conspecifics in order to maintain certain roles from which relationships may emerge (Whitehead 2008). Signals in a variety of modalities may be sufficiently variable to enable discrimination and identification. Bewick's swans have unique faces (Scott 1988), individuals in several deer species have unique olfactory secretions (Lawson et al. 2000), and a variety of species produce individually distinctive vocalizations (e.g., McGregor & Peake 1998). In both olfactory (e.g., Palagi & Dapporto 2006) and acoustic modalities (e.g., Cheney & Seyfarth 1980; Blumstein & Daniel 2004), individually distinctive signals (**Figure 8.6**) may be used for "true" individual recogni-

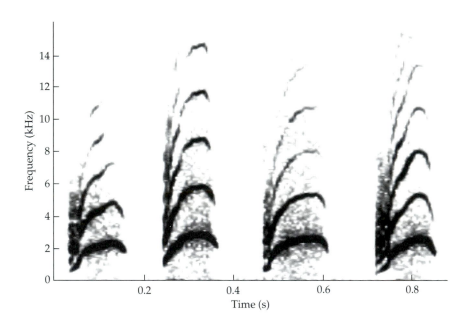

FIGURE 8.6   Individually distinctive alarm calls from Olympic marmots (*Marmota olympus*). The spectrograms are recordings of the alarm calls of four different Olympic marmots. Note the differences in frequency contours and the presence and extent of the initial pulse of sound. (Marmot recordings by Sue Griffin.)

tion; that is, it is reasonable to infer that the signals create a representation (Evans 1997) in the perceiver.

Social species may benefit from having unique vocalizations, and the vocalizations of more-social species may contain more information about the identity of the sender than those of less-social species (e.g., Beecher 1989b; Pollard & Blumstein 2010). If little is known about a rare species, the amount of information contained in its vocalizations may be a useful index of the relative degree of sociality. Specifically, since sociality is associated with the evolution of individuality, it should be possible to infer historic patterns of sociality from vocalizations or other individualistic signals. And since knowledge of social behavior may have important implications for population management (see Chapter 10), knowledge of the degree of individuality may help population management for species in which little is naturally known about their sociality.

In addition to the potential that vocal diversity can help to reconstruct the degree of sociality of a rare species, these individually distinctive vocalizations can be used as marks of identity. Normally, individuals must be marked and recaptured to estimate population sizes. However, if known individuals can be recorded at one time and then rerecorded at another time, the population size can be estimated in this way as well (McGregor & Peake 1998).

How to Quantify Species-Specific and Individually Specific Vocalizations

There are two complementary problems: (1) determining whether vocalizations are species-specific or individually specific and (2) assigning recorded calls to specific species or individuals. We will discuss both in order. For this discussion, we will assume that some systematic sampling regime was used to obtain a library of calls from different individuals in one or more species. For the first problem, we must know the identity of the sender.

Describing species specificity or individuality

For us to determine whether an acoustic signal is species-specific, species A should typically sound like species A, while species B should typically sound like species B. Similarly, if an acoustic signal is individually specific, individual A should typically sound like individual A, while individual B should typically sound like individual B, etc. Statistically, this means that within-species or within-individual acoustic variation must be less than between-species or between-individual variation. To determine whether this is the case, you must first have multiple recordings from a species or an individual. These recordings must come from different occasions. Thus, recording ten songs in a row from a single bout of song is not ideal, because there may be some external stimulus that reduces (or increases) variation. Considerable acoustic variation can be explained by environmental features (see above) and the social situation. For instance, when you say "I'm going to light a fire in the fireplace," the word *fire* sounds very different than if you were to say, while in a crowded theater, "FIRE!" Another reason to obtain multiple recordings from the same individual at different times is that they will be more statistically independent.

Once you have obtained recordings from different occasions, you must measure them. Deciding what to quantify requires a mix of intuition and experience. Because individuality is often communicated by varying frequency characteristics (e.g., Leger et al. 1984; Conner 1985; Blumstein & Munos 2005), you should be certain to measure several (rather than one) parameter associated with the frequency structure of the vocalizations (see Box 8.1).

There are several methods of classifying calls of individuals. We will mention a relatively straightforward one. **Discriminant function analysis (DFA)** is a statistical technique that creates equations to assign group membership from a matrix of measurements made on two or more groups. In the case of discriminating among individuals, you must create a matrix of acoustic measurements. Ideally, you will have the same number of observations for each individual (but this is not necessary), and the DFA will determine which variables are most important for individual discrimination and will use the discriminant functions to classify individuals into their preassigned groups.

Assume individuals sing very distinctive songs. For instance, one individual sings a song that sounds like *A*, another *E*, and another *I*. Also assume that there is some variation in how a given individual sings its song. Thus, there will be some within-individual (also called within-class) variation (i.e., sometimes *A* sounds like *hey*, and other times it sounds like *aaii*). As long as the between-individual variation is greater, it should be possible to distinguish individuals, and your matrix of acoustic measurements will enable you to assign identity relatively easily. One statistic from the DFA that can be calculated is the percent correct classification. When you have high percent correct classifications, the DFA can assign calls to callers with high confidence. Thus, DFA can help determine whether it is appropriate to use vocalizations to census individuals (McGregor et al. 2000). An example of how DFA classifies cases can be found at www.statsoft.com/textbook (then go to Discriminant Analysis).

Assigning recorded calls to individuals

If you have multiple recordings of all individuals in your population, you can calculate discriminant functions and use them to predict identity of an unknown caller, given a series of measurements on that unknown caller. Having a sufficient set of recordings for all individuals may or may not be likely, depending on your system and question of interest (McGregor et al. 2000). For instance, in a semicaptive situation, you might have recordings of all individuals. In such a case, you might use the acoustic census technique to avoid capture-related stress while ensuring that animals were still present. By contrast, in the wild, it is unlikely that you would have recordings from all individuals in a population.

There are techniques that compare the similarity of two (or more) recordings. **Spectrogram correlation** (also known as spectrogram cross-correlation) systematically shifts two spectrograms along their *x*-axes (in the bioacoustic programs Canary and Raven) or along both *x*- and *y*-axes (in the commercially available bioacoustic program SIGNAL) to identify the maximum similarity between them. Spectrogram correlation values range from 0 to 1.0, where 1.0 indicates an identical signal. It is relatively easy to generate a matrix of maximum similarity values, but it is relatively difficult to determine the threshold similarity value that would assign individual calls to the same or different classes. This is no different from other sorts of discrimination/classification problems where you trade off error with certainty (e.g., McGregor & Peake 1998; McGregor et al. 2000). Nevertheless, a combination of DFA and spectrogram correlation should allow you to set reasonable similarity thresholds to identify and classify calls from unknown callers.

Assuming there are individually specific calls, how might one calculate the information about caller identity in a call? One way is based on information theory (Beecher 1989a). To use this method, measure a number of acoustic characteristics (see Box 8.1). The actual parameters measured may vary with the species, but the goal is to identify those that are variable. Then, conduct a principal components analysis to reduce the set of variables into a set of uncorrelated variables (see Chapter 9, Box 9.1). Next, fit an ANOVA, grouped by individual to those principal component scores that are significant (i.e., $p < 0.05$). To calculate the statistic H_i, the amount of information about an individual contained in trait i, you will need the F-statistic from the ANOVA (F) and the number of individual samples (n). Use the following equation.

$$H_i = \log_2 \sqrt{\frac{F + n - 1}{n}}$$

Calculate this for each trait (i.e., each component score), and sum across traits (i.e., component score). The sum of the H_i, across all traits, H_s, is the information about individuals contained in the set of measured traits. Species with larger H_s scores have more individually specific information contained in their calls.

Application: Using Individuality to Estimate Population Size of Endangered/Threatened Species

McGregor & Peake (1998) and McGregor et al. (2000) describe how they used individual vocalizations to noninvasively census corncrakes (*Crex crex*), a threatened European nocturnally active rail. They first determined from a set of known callers that calls were individually distinctive, and they identified a feature—the duration of a pulse of sound—that was particularly useful in discriminating among callers. They then used spectrogram correlation as an index of similarity and explored different thresholds that would allow reasonable trade-offs of certainty of assigning an unknown call to an individual with error. By using this information about the identity of individual callers, and by employing "mark-recapture" algorithms, they were able to generate a better estimate of the true population than that which was possible by capture-recapture techniques.

Fernández-Juricic et al. (2009b) studied the potential of using individuality in the vocalizations of the US-endangered southwestern willow flycatcher (*Empidonax traillii extimus*) in acoustic surveys. The study found that one important confounding factor in assigning vocalizations to individuals is ambient noise, which can be reduced by recording individuals across shorter distances. Furthermore, the researchers recommended using a combination of discriminant function analysis and artificial neural networks to identify individuals; the former performed better at discriminating between individuals, and the latter at identifying new individuals in the population. Artificial neural networks differ from discriminant function analysis in the computational approach used to assign individuals to a particular class.

In addition to DFA and neural network algorithms, other approaches (beyond the scope of this book) include using Bayesian classifiers (Hayward 1997), hidden Markov models, and fuzzy logic (Kirschel et al.

2009b) to classify individuals. In contrast to DFA, all other techniques require you to have some degree of computer programming ability to get them to work.

Further Reading

Baptista & Gaunt (1997), McGregor & Peak (1998), and McGregor et al. (2000) discuss a variety of bioacoustic techniques that may be useful. Topics we did not cover include using playback to census individuals (described in Bibby et al. 2000) and using vocalizations as a taxonomic tool (described in McCracken & Sheldon 1997). Rabin et al. (2003) and Warren et al. (2006) are good reviews of the effects of anthropogenic noise on animal behavior. In addition to Canary (www.birds.cornell.edu/brp/software/canary-information) and Raven (www.birds.cornell.edu/brp/raven/RavenOverview.html), Praat (www.fon.hum.uva.nl/praat) and SoundRuler (http://soundruler.sourceforge.net/main) are freely available, cross-platform general purpose acoustic packages. The freely available package Ishmael (www.pmel.noaa.gov/vents/acoustics/whales/ishmael) has some recognition algorithms built in.

9 Individuality and Personalities

As anyone with a dog or cat realizes, nonhumans have distinctive personalities (Gosling 2001; Jones & Gosling 2005). The *Cambridge Dictionary of American English* defines *personality* as "the special combination of qualities in a person that makes them different from others, as shown by the way they behave, feel, and think." We use the term *personality* not to be anthropomorphic but rather to point out that nonhumans may behave predictably and consistently in different situations, much as human personality implies that you may behave predictably and consistently in different situations (Gosling 2001). From a theoretical perspective, it is interesting to understand whether this behavioral variation is adaptive, and what maintains it (e.g., Sih et al. 2004a,b). For instance, being cocky might be costly in certain situations (you might not want to be too cocky if you are at a bank during a robbery!) and beneficial in others (cocky men might be more likely to meet new women in loud bars). Similarly, nonhuman personality types may have distinct costs and benefits associated with them, and thus personality may have fitness consequences (Réale et al. 2007). Wildlife biologists armed with an understanding of the fitness consequences of personality types may be able to use this knowledge to increase captive-breeding success and the success of reintroductions or translocations for conservation if and when individuals with certain personality types are easier to breed or more likely to survive upon release.

What Is Personality?

This is an exciting time in behavioral ecology because after years of mostly treating individual variation as statistical noise, researchers have begun to focus specifically on individuality and to ask questions about how it persists and how we can understand its evolution. We will distinguish personality traits, or temperament, from behavioral syndromes. **Personality traits** are consistent individual differences over time (Réale et al. 2007). **Behavioral syndromes** are correlations of behaviors or personality traits across different contexts (Sih et al. 2004a,b). Both personality traits and behavioral syndromes have implications for conservation and management, because they tell us something important about how animals interact with their environments (e.g., they may be shy or bold, active, aggressive) and because they may have fitness consequences.

In a wide-ranging review, Réale et al. (2007) defined five different types of temperament traits and described how to measure each of them:

- *Shyness–boldness*, which is determined by how individuals respond to risky, but not necessarily novel, situations
- *Exploration–avoidance*, which is determined by how individuals respond to a novel situation
- *Activity*, which is defined as how much individuals move around in riskless and nonnovel situations
- *Aggressiveness*, which is determined by how individuals react aggressively toward conspecifics
- *Sociability*, which is determined by how individuals react amicably toward conspecifics

We believe that these traits, and their measurement, constitute an efficient and consistent way to describe behavioral differences among individuals. By understanding them, we may be able to increase captive-breeding success and the success of reintroductions and translocations for conservation.

Quantifying Personality Types

Given that animals naturally do many different things, how do we quantify personality types? Three different techniques are routinely used to quantify personality: observations of experimentally induced behavior, expert evaluations by those familiar with the individuals (i.e., keeper surveys), and observations of natural behavior (Gosling 2001). All three methods generate a number of behavioral measures (e.g., time budgets of different activities, how animals behave in different circumstances) that are subsequently analyzed using principal components analysis (**Box 9.1**) to enable factors (dimensions) to be extracted. In some cases, only the first factor is interpreted; in others, more than a single factor is interpreted.

Observations of experimentally induced behavior

Mirror image stimulation (MIS) is a technique that has been used by psychologists and behavioral ecologists to identify personality types. An animal is placed in an enclosure with a mirror, and its behavior is recorded. In theory, it is possible to place mirrors in less controlled situations and bait animals to them to see how they interact with a mirror, and thus MIS can be

BOX 9.1 Principal Components Analysis

Principal components analysis (PCA), a type of factor analysis, is a statistical technique designed to take a set of somewhat correlated variables and reduce their dimensionality so as to allow the underlying structure of those variables to be identified. In other words, PCA has the potential to reduce and synthesize the number of variables that characterize a biological system. Wildlife biologists may be most familiar with principal components analysis as it is applied to quantifying body size or vegetation structure. Say we make a series of measurements on a sample of birds. For each bird, we measure the length and width of the wing and tail, and we measure body mass. Ultimately, we might wish to develop one measure that describes body size. Wing length and width, tail length and width, and body mass are all ways to think about body size, but for some subsequent statistical analysis, we might need a single metric summarizing all these parameters. What to do?

PCA can be used to reduce the dimensionality of these data. It does so by identifying correlated variables and creating equations that define those correlated variables. For instance, if we were only concerned with the wing measurements, we would see that wing width is correlated with wing length within a species, and the regression would give us a predictive equation. We are all familiar with the equation that predicts any y-value from a line: $y = mx + b$, where m is the slope of the line and b is the y-intercept. PCA gives us similar sorts of prediction equations, except that we are trying to predict an individual's component score based on a set of measured variables and estimated parameter values.

In a two-variable situation, we would extract one factor (i.e., we would have a regression equation that explains the relationship between two variables). In the multivariate case (we have 5 different variables representing body size), we can extract as many as $n - 1$ (in this case, 4) factors. The actual number of factors extracted will depend on the correlations among them. In the example above, it is possible that we might extract 3 factors: one describing characteristics of the tail, another describing the wing, and the third body mass. The predictive equations for a given factor have coefficients, much like the m and b values from the equation for a line, that allow us to predict an individual's factor score given its wing length, wing width, tail length, tail width, and body mass.

When quantifying personality types, we use potentially correlated measures of time allocation, rather than body size measurements, and we interpret the factors extracted by scrutinizing each factor's coefficients. By tradition, coefficients $\geq |0.7|$ are interpreted as "loading highly" on a given factor and help define the factor. We provide an example of PCA in the text.

used in the field. According to Réale et al. (2007), MIS can be used to identify aggressiveness and sociability, although previous researchers have interpreted results from MIS as illustrating a shyness–boldness trait (Svendsen & Armitage 1973).

Blumstein et al. (2006) used MIS to quantify the behaviors of the endangered Vancouver Island marmot (*Marmota vancouverensis*). They placed a mirror in a marmot enclosure and waited for a marmot to discover the mirror. They then used JWatcher to score the videotape for the 30 seconds before the marmot's initial discovery of the mirror and for the 10 minutes following the discovery. They quantified vigilance, foraging, interactions with the mirror, locomotion, tail-flagging behavior (a behavior that illustrates "self-confidence"), and escape into the burrow. They then used PCA to extract fewer factors that would characterize the system. They found that some marmots initially investigated the mirror, while others initially fled. The first factor in the PCA was interpreted as an interactive factor, while the second factor was interpreted as a fearful factor. It was found that animals that were responsive were not that vigilant, interacted a lot with the mirror, and foraged a lot. Those that were fearful spent a lot of time in their burrows, moved around a lot, and tail flagged.

Because Blumstein et al. (2006) were interested in searching for syndromes, they repeated the mirror experiment a second time, and for this second MIS, they placed a taxidermic mount of a wolf in the enclosure. They extracted principal components for the MIS-wolf experiment and then correlated factor scores across these two situations. In both cases, the first factor extracted was interpreted as an interactive factor—what Réale et al. (2007) would refer to as sociability. The extracted factors explained about 72% of the variation in responsiveness, and the factors calculated in each of the two experiments were correlated with each other. Because these responses were correlated, this pattern was identified as a syndrome.

A parallel set of experiments with the same animals demonstrated that how marmots respond to the MIS predicts the magnitude of response to a key predator. Because quantifying predator recognition is much more difficult (see Chapter 7), the MIS experiment here discovered a way to indirectly measure predator recognition abilities. In terms of fitness consequences of this syndrome, some of these animals were subsequently released, and the male that was identified in the study as the most interactive was killed by a cougar.

In addition to using mirror image stimulation, researchers often expose animals to novel environments and quantify their responses. For example, work with great tits, a species that may nest in nest boxes, relied on removing animals from their nest boxes for up to 24 hours, bringing them into captivity, and quantifying how they responded to a novel environment (e.g., Both et al. 2005). In these sorts of studies, factors explaining the response to the novel environment are used to understand ecologically important questions (natal dispersal, dominance, survival, or mate choice).

Similar work with bighorn sheep (*Ovis canadensis*) by Réale and colleagues relied on catching free-living animals (Réale et al. 2000; Réale & Festa-Bianchet 2003). Individuals differed in the ease with which they were trapped, and this was interpreted as an indication of boldness. Importantly, once the sheep were trapped, the researchers quantified the ease with which they could be handled. Docility measures were repeatable and heritable. There was selection on bolder ewes to reproduce earlier than shier ewes, and more docile ewes reproduced earlier than less docile individuals (**Figure 9.1**). Since age at first reproduction has a profound effect on lifetime reproductive success, these personality traits have important fitness consequences.

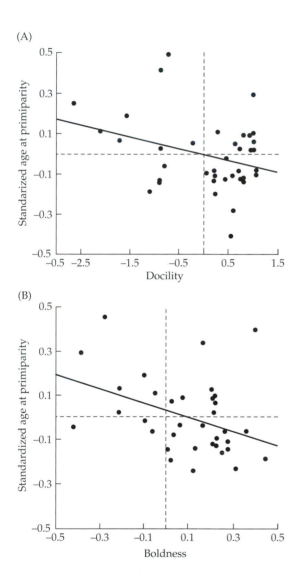

FIGURE 9.1   (A) More docile and (B) bolder female bighorn sheep first reproduce at earlier ages. Primiparity is expressed here as the residual from a linear model that includes population density, body mass, and either boldness or docility, depending on the analysis. Since age at first reproduction has important demographic consequences, variation in these repeatable personality types has important demographic consequences. (After Réale et al. 2000.)

FIGURE 9.2 Bolder swift foxes were more likely to die in the first six months following release than less bold foxes, demonstrating a survival cost to boldness. (After Bremner-Harrison et al. 2004.)

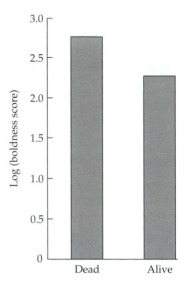

Bremner-Harrison et al. (2004) discovered that shier captive-bred swift foxes (*Vulpes velox*) survived longer following reintroduction from a captive-breeding program than did bold animals (**Figure 9.2**). She quantified shyness and boldness by exposing captive animals to a novel object (a beach ball or a bag stuffed with crumpled paper that made noise when moved) and used PCA to reduce the observed behaviors into factors. Postrelease monitoring generated the survival data set. Fox survival upon release was a function of personality type. Thus, personality has fitness consequences in the context of reintroduction for conservation.

Keeper surveys

People who work with animals know those animals well. Surveying keepers can be a robust method of quantifying personality types in captive individuals. For instance, Gosling (1998) asked four caretakers of a population of 34 spotted hyenas (*Crocuta crocuta*) to rate individual hyenas on 42 different traits. When he subjected the traits to a PCA, he extracted five dimensions that he called assertiveness, excitability, human-directed agreeableness, sociability, and curiosity. These five factors explained 78% of the total variance.

Carlstead et al. (1999) sought to identify causes of breeding failure in captive black rhinos (*Diceros bicornis*). They collected data from 23 zoos and surveyed the rhinos' keepers to understand rhino behavior as a function of the captive environment. They extracted six behavioral traits: olfactory behaviors, chasing/stereotypy/mouthing, fear, friendliness to the keeper, dominance with conspecifics, and patrolling. They found that fearful rhinos were housed in conditions with greater public access, and they recommended reducing public access to reduce the stress on male rhinos. Males that scored highly on the dominance factor were those in small enclosures. Females that scored highly on chasing/stereotypy/mouthing (a metric of agitation) were housed in smaller enclosures. Importantly, the researchers found that some of these factors were associated with reduced breeding success.

Observations of natural behaviors

Determining personality traits or syndromes from studies of free-living animals is more difficult because many uncontrolled factors may explain variation in a particular variable (Réale et al. 2007). For instance, how much time an individual allocates to foraging can be influenced by both competition with conspecifics and predation risk (Elgar 1989). The ability to escape a particular area may be influenced by the incline or substrate that an animal must run over (Blumstein 1992). Depending upon how your data are collected, either there may be systematic confounding factors or there may just be obscuring variation that makes it difficult to identify personality traits or syndromes.

In a study of free-living yellow-bellied marmots, Blumstein et al. (2004b) correlated residuals from two linear models (one studying locomotor ability, the second studying antipredator vigilance) to identify a "locomotor ability–wariness" behavioral syndrome. (Similar techniques can be used to identify personality traits.) Specifically, they first modeled maximum running speed by studying factors that influenced run time as a function of substrate, incline, body mass, and distance run. They then modeled time allocated to foraging/vigilance as a function of age–sex, incline, group size, date, and substrate. After variation explained by a variety of potentially important extrinsic factors was controlled for, the investigators suggested, the residuals from these linear models should reflect the "intrinsic" speed and wariness of subjects. There was modest repeatability of vigilance residuals, suggesting that after extrinsic factors that explained vigilance were controlled for, individuals differed in their degree of vigilance. Blumstein et al. found that running speed residuals were correlated with vigilance residuals in a way that suggested that slow marmots minimized their time exposed to predators while foraging (slow marmots looked the least and foraged the most during foraging bouts; **Figure 9.3**). More recent work (Blumstein et al. 2010) demonstrates a heritable genetic basis for these traits and a weak genetic correlation between them.

Determining consistency and repeatability

If animals have stable personalities, we should be able to reliably measure these traits. This means that our measurements should be repeatable. In some cases, it may be desirable to measure an individual only once in a particular situation (e.g., when using a predator model, so as to not habituate the individual to predatory stimuli), but whenever it is possible, we suggest

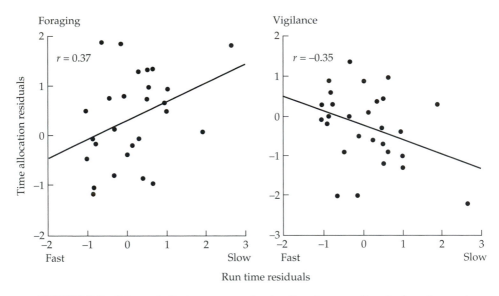

FIGURE 9.3   Marmots that run more slowly allocate more time to foraging and less to vigilance than those that run more quickly. Such a phenotypic correlation suggests that there is a locomotor ability–wariness syndrome. (After Blumstein et al. 2004b.)

measuring individuals multiple times. By calculating the repeatability of a trait (Boake 1989), one can see how consistent individuals are.

There are several ways to calculate **repeatability**; all require at least two observations per individual. First, fit an ANOVA model with your variable of interest as the dependent variable and the individual as a fixed independent factor (Falconer & Mackay 1996). Second, fit a mixed effects model with your variable of interest as the dependent variable and where individuals are entered as an independent random effect (Díaz-Uriarte 2002). Third, calculate the within-class correlation coefficient (Sokal & Rohlf 1981) from individuals observed multiple times. Because you interpret the within-class correlation coefficient as you would a normal correlation, these results will tell you approximately how much variation is explained by individual differences. If there are significant effects of the individual, or if the within-class correlation coefficient is substantial, then there is a high degree of repeatability.

Repeatability is important for another reason; it may set an upper limit on a trait's heritability (Boake 1989). **Heritability** is calculated by dividing the additive genetic variation (V_a) by the total phenotypic variation ($V_a + V_e$, where V_e is the environmental component of phenotypic variation). Heritability estimates allow us to infer past selection on traits (which may reduce additive genetic variance) and shed insight into the future evolutionary potential (traits can evolve only if there is significant additive genetic variation). The heritability of personality traits has been studied in the lab, via breeding experiments (van Oers et al. 2004; Sinn et al. 2006). Importantly, heritability can be directly studied in the field. To do so, we must have a genealogy and then use the "animal model" (a statistical technique, beyond the scope of this *Primer*, that allows you to estimate additive genetic variation based on field measurements; Kruuk 2004) to estimate the heritability of a trait (behavioral, morphological, life history, etc.). However, for systems without these hard-earned genealogical data, we are left with just estimates of repeatability. Repeatabilities should be used with some caution (Dohm 2002). However, high repeatabilities tell us that there might be evolutionary potential because there is likely to be heritable variation upon which selection may act. By contrast, low repeatabilities tell us that there is no further opportunity for a personality trait to evolve. Lack of heritable variation in personality types might constrain a species' ability to expand into new habitats, or it might make a population particularly vulnerable to habitat changes.

Should We View Personality Dimensions as Independent, or Should We Look for Syndromes?

Given a set of quantitative (or qualitative) observations of a set of individuals, you can use principal components analysis to extract factors. Those factors describe personality dimensions. The analyses can be conducted in different contexts (e.g., mating contexts, foraging contexts, aggressive contexts). If either the raw data (e.g., proportion of time foraging in a particular situation) or the factor scores (e.g., a boldness factor score) are correlated across different contexts, then a behavioral syndrome has been identified (**Figure 9.4**).

There is some controversy over whether personality traits are independent or correlate with each other, creating syndromes, and there is empirical support for both patterns (Réale et al. 2007). For instance, selection may favor the evolution of independent temperament traits, each in response to a specific context (Wilson 1998). But evidence does exist that there are correlations among traits across contexts that create trade-offs (Sih et al. 2004a).

At one level, the interest in syndromes parallels the interest in personality types: if individuals behave predictably, it is essential to understand whether they do so across contexts. Both personality types and syndromes may have fitness consequences, and both may be heritable. To better understand performance in one situation or context, it is important to think about the effects of a behavior in other situations or contexts. For instance, if aggressive behavior in one context carries over to parental care, then to understand parental care, you really need to understand aggressive behavior, and vice versa.

We have written about antipredator behavior (see Chapter 7) because we should develop a fundamental understanding of this type of behavior in

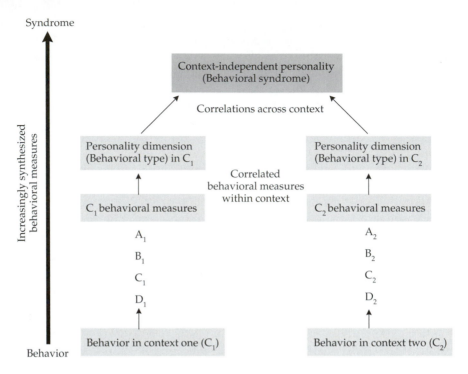

FIGURE 9.4   The distinction between a personality type and a syndrome. Assume you measure a number of behaviors (e.g., time allocated to vigilance, foraging, grooming) in two contexts (e.g., with or without a predator present). Correlated measures within a context define a personality dimension (e.g., shy, bold). Correlations across contexts define a behavioral syndrome. (After Smith & Blumstein 2008.)

species that need to be reintroduced or translocated to predator-rich locations. In the context of antipredator behavior, people have talked a lot about shy–bold or proactive–reactive syndromes. In some cases, animals are systematically bold in novel situations. And there may be consequences of being bold. For instance, animals bold around predators may also be very aggressive, as has been shown in sticklebacks (Huntingford 1976). These syndromes may be adaptive (if being bold around predators and aggressive is favored), or they may illustrate constraints on adaptive evolution (it might be important to be aggressive to conspecifics but detrimental to be bold around predators). Regardless, to fully understand antipredator behavior, we need to understand personality types and syndromes.

Fitness Consequences of Personality

In a formal meta-analysis (see Chapter 3), Smith & Blumstein (2008) reviewed published studies reporting fitness consequences of personality types to identify general trends across species. They focused on boldness (the willingness of individuals to take risks), aggression (responses of individuals to conspecifics), and exploration (movement of animals in novel situations or locations) and looked at two fitness correlates—survival and reproductive success. They found that bolder individuals, particularly males, had higher reproductive success but incurred a survival cost, thus supporting the hypothesis that variation in boldness was maintained due to a trade-off in fitness consequences across contexts. While there was variation in both exploration and aggression, Smith & Blumstein found only a small positive effect of exploration on survival, and there was a positive effect of aggression on reproductive success but not on survival. In some cases, captive-reared animals behaved differently than wild-caught animals, although there were insufficient data to properly determine whether captive breeding within a species differentially influenced personality traits or subsequent survival.

Applications

Personality traits are typically thought of as being continuous. Thus, within a personality dimension, there are shier and bolder individuals and there are more and less sociable individuals. Because these traits may affect both reproduction and survival, it is essential to consider them when breeding animals in captivity or when planning a reintroduction.

Studies, such as those by Carlstead et al. (1999), which focused on correlates of breeding in rhinos, and Bremner-Harrison et al. (2004), which studied correlates of survival following release in swift foxes, illustrate how such personality and syndrome data may be used to manage captive animals and increase reintroduction success. Watters & Meehan (2007) argue that in nature animals have different skills and personalities and that paying more attention to this phenotypic variation is essential for increasing postrelease survival.

A defining characteristic of such studies is that animals must be systematically screened before they are paired, or before they are released. By screening, we develop both a data set useful for adaptive management and a tool that can help determine which animals might be suitably paired or released at a particular time.

However, these studies also raise an interesting ethical question: What happens if one behavioral type prospers in a captive environment? Or what happens if one behavioral type survives better following reintroduction? In both of these cases, we have implicitly selected for certain behavioral types, and we have reduced the natural variation in the environment. If this variation is adaptively maintained (Smith & Blumstein 2008; Watters & Meehan 2007), then we have changed a population's behavioral variation. Buchholz & Clemmons (1997) pointed out that local adaptations maintain behavioral variation, and maintaining behavioral variation is an important part of wildlife conservation. Thus, at some very important level, if we lose natural variation, we may lose something important in our attempt to recover a threatened or endangered species.

These possible scenarios succinctly illustrate the frustration of single-species management: when we begin to fiddle with systems we only superficially understand, there may be unintended consequences. That said, with a fundamental understanding of behavioral variation, it should be possible to maintain it. For instance, if individuals with certain personality types breed better than others, they can be used to rapidly increase a captive population size. However, efforts to breed those personality types that are difficult to breed should continue so as to maintain behavioral variation. And if some individuals fare better when initially reintroduced to the wild, these should form the founder populations, while other individuals should be maintained in captivity and released later. Thus, recognizing the potential problem of maintaining behavioral variation allows us to solve it.

Further Reading

Gosling (2001), Sih et al. (2004a,b), and Réale et al. (2007) are required reading for developing a fundamental understanding of why personality and syndromes are important and how they can be studied.

10 Demographic Consequences of Sociality

A population's or species' social structure, social interactions, or social behavior may affect its persistence. We will discuss in Chapter 11 the demographic consequences of sexual selection and mate choice, but in this chapter we will discuss other demographic consequences of sociality. Features such as conspecific attraction may generate insights into how to attract animals to particular locations. But conspecific attraction may also make a species more vulnerable to exploitation by hunters or anglers. Other features, such as grouping, may reduce predation risk and thus have demographic consequences. And, because social structure and social behavior are often adaptive responses to environmental variation, anthropogenic changes may have a profound influence on social behavior and, consequently, demography and population viability. In this chapter, we will describe these and other potential links between social behavior and wildlife conservation and provide examples to illustrate them, with the hope of inspiring the development of potentially novel wildlife management tools.

What Is Sociality?

Sociality has many attributes and thus means different things to different people. One attribute of sociality is how many animals live together. We might feel comfortable saying that a colonial group of swallows is more social than a species in which mated pairs defend territories. However, what goes on in a colony may cause some reflection: What if pairs within a colony defend nesting sites and, for all intents and purposes, do not interact much with other members? What if colonial animals are constrained to be colonial and are more stressed and have more diseases? What if individual reproductive success is lower in colonies than when animals live alone?

Another way to explain social variation is to focus on mating systems, which may vary both inter- and intraspecifically (Lott 1991; Alcock 2009). Mating systems are named with respect to the number of males a female mates with and the number of females a male mates with. In a **monogamous** system, any female or male mates with only a single individual. Demonstrating genetic monogamy often requires the use of molecular parentage assignment, because individuals may be socially monogamous but engage in copulations outside of the pair. Social monogamy may be stable; some long-lived birds each pair with only a single mate. Or social monogamy may be serial, in which case pairs are stable within breeding seasons but pairings may change between seasons. In a **polyandrous** system, a female mates with more than one male. In a **polygynous** system, a male mates with more than one female. In **promiscuous** systems, both males and females have multiple mates. Because the number of mates has a strong influence on genetic variation, it is important to understand the genetic and demographic consequences of mating systems.

From another perspective, more-social species have more well-defined social roles (Hinde 1975). Roles are things that animals do. For instance, there may be breeders and helpers, individuals that specialize in defense, and others that specialize in food acquisition. In these cases, sociality is not simply having a lot of animals in one location; rather, it means more about what specifically the animals are doing in those locations.

Relatedness matters too. Contrast two groups of equal size: one composed of relatives and one composed of unrelated individuals. We expect different social dynamics in the one composed of relatives, because individuals can obtain both direct and indirect fitness from their actions.

What are direct and indirect fitness? Successful individuals are those whose genes will be present in the future. There are two ways to achieve this. First, you can do things that allow you to produce a lot of descendants. As we will discuss in Chapter 11, individuals vary in terms of how many offspring they leave. The number of offspring produced is a measure of **direct fitness**, and when we think of ways to have our genes outlive us, producing offspring makes a lot of sense. However, sometimes animals do things that apparently benefit others. If those others are relatives, J. B. S. Haldane's famous quip about how he would lay down his life to save "two brothers or eight cousins" applies. By helping relatives reproduce, you will leave genes indirectly; this is a measure of **indirect fitness**. Specifically, an individual shares 50% of its genes with a full sibling and 12.5% with a first cousin—thus, all the individual's genes could persist if it helped save two brothers ($2 \times 0.5 = 1.0$) or eight first cousins ($8 \times 0.125 = 1.0$). By thinking about ways that animals obtain indirect fitness by engaging in potentially costly activities, we can often explain altruistic behavior. The fact that animals can obtain fitness both directly and indirectly has many consequences and implications for social evolution, and some of these have demographic implications.

Thus, if faced with managing social behavior, we should consider a variety of things. We will discuss several with potentially large effects. The key question to ask oneself when faced with a potential social behavior management problem is: What is the relationship between social behavior trait X and demography? Practically, this asks whether social behavior X has fitness consequences, and what they are.

Reproductive Skew

Relatedness among group members is likely to influence cooperative and competitive social behavior, and these are likely to influence reproduction. Consider, for instance, reproductive skew. Unequal sharing of reproduction—**reproductive skew**—within groups is a pervasive phenomenon (Hagar & Jones 2009). Skew is a consequence of which individuals breed, thus anything that influences which ones reproduce can have effects on gene dynamics and, therefore, the effective population size, N_e (see Box 11.1). Species like elephant seals have one or a few males on a beach monopolizing access to virtually all the females (**Figure 10.1**). The term skew means that a few individuals obtain most of the matings while most have no reproductive success.

Ecological constraints have long been known to be important factors explaining social system variation (Lott 1991). For instance, cooperatively breeding and reproductively suppressed Seychelles warblers that were translocated to an isolated island without resident warblers suddenly start-

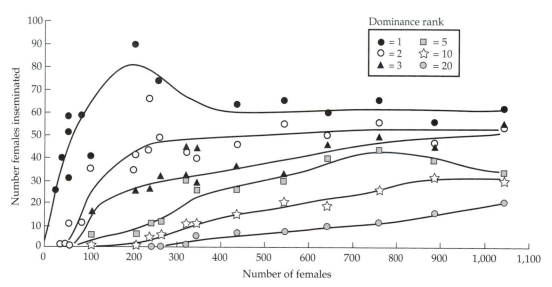

FIGURE 10.1   Reproductive success of male elephant seals as a function of dominance rank and the number of females on the beach. (After Le Boeuf & Reiter 1988.)

ed reproducing. As the small island began to be saturated with breeding territories, newly born residents became reproductively suppressed (Komdeur 1992). Similarly, the removal of resident breeding male superb fairy-wrens (*Malurus cyaneus*) led to the immigration of new males that were previously helping their parents raise young in their natal territories (Pruett-Jones & Lewis 1990). In both cases, the lack of available breeding vacancies was identified as an ecological constraint on breeding that generated reproductive skew.

Patchy environments limit dispersal ability and form another possible constraint implicated in the origin of reproductive skew. For instance, in the family of rodents that includes the eusocial naked mole rat (*Hetercephalus glaber*), widely scattered food, arid habitats, and hard soils are hypothesized to select for group living (Lacey 2000). **Eusociality** is a breeding system characterized by a solitary breeding female and helpers that has evolved to such a degree that there are morphological adaptations to increase the ability of the breeder to produce more offspring. In the case of the naked mole rat, breeding females have extra vertebrae that increase their size and allow them to grow more young. Interspecific variation in mole rat group size is associated with food density and rainfall (Faulkes et al. 1997).

We should generally expect individuals to compete for reproductive opportunities, groups to be characterized by skew, and the genetic consequences of this skew to have consequences for population persistence through its effects on N_e. Identifying the causes and consequences of skew within a species may provide strategies by which managers can increase the likelihood that a population will persist over time. For instance, reducing skew, by reducing ecological constraints, should increase N_e and enhance population persistence.

Reproductive Conflict

Complex sociality is characterized by a reduced probability that all individuals will reproduce (Blumstein & Armitage 1998). Often this condition is established through **reproductive suppression**, whereby potentially fertile females do not breed (Solomon & French 1997). If females living closely with others compete reproductively, then not all females breed, or litter sizes are reduced. Reproductive suppression therefore reduces the effective population size by reducing the number of potentially reproductive females who are able to breed (Anthony & Blumstein 2000).

Habitat saturation may be a direct cause of reproductive suppression and **alloparental helping**, that is, nonparents helping to rear the offspring of other group members. Some species, when faced with no chance to breed independently, engage in alloparental care and therefore may obtain indirect fitness benefits by helping their relatives rear young. Such helping behavior has been demonstrated to be a strategy whereby individuals make the "best of a bad job" because, as discussed above, when individuals are translocated to a location with potentially available territories, they immediately begin breeding independently (e.g., Komdeur 1992). Thus, identifying species in which habitat saturation suppresses independent reproduction gives us a tool to use, should managers need to increase the number of breeders. For instance, for red-cockaded woodpeckers (*Picoides borealis*), Walters et al. (1992) provided new cavities (by drilling holes in trees), and this allowed formerly nonbreeding but reproductively mature individuals to breed.

Alberts et al. (2002) experimentally removed dominant male Cuban iguanas (*Cyclura nubila*) from a population where they prevented subordinates from breeding. This removal led to formerly subordinate males taking over vacated territories and winning more fights. While the researchers were unable to directly study the effect on reproductive success, such interventions might help increase the number of breeding individuals and thus, they argued, prevent the loss of genetic variation.

Mechanisms of suppression can be sophisticated and may also affect captive animals that, in theory, have sufficient resources. One mechanism of suppression is via social stress (Wingfield & Sapolsky 2003). Such stress-induced sterility works through both the hypothalamic–pituitary–adrenal (HPA) axis and the hypothalamic–pituitary–gonadal (HPG) axis. Life history theory leads us to expect that reproduction is traded off against growth and maintenance. Thus, when animals are particularly stressed, they should allocate energy away from reproduction and growth and mobilize energy to the systems used in escaping risks (in this case, threatening conspecifics). It is this reallocation of energy that leads to stress-induced sterility. If social animals, in which reproductive suppression is known or suspected to occur in the wild, fail to breed in captivity, then a strategy to increase reproduction might be to reduce the opportunity for social stress by moving animals apart. However, in some species, social facilitation is required for breeding (Hearne et al. 1996; Swaisgood et al. 2006). It is therefore essential to conduct formal experiments to see whether stress-induced reproductive suppression is reduced by moving potential breeders apart or whether grouping facilitates reproduction.

Social Behavior Reduces Mortality

In some cases, it is safer to be in a crowd than alone. Several models of predation hazard assessment show that per capita risk declines as group size increases (Krause & Ruxton 2002). This may result from a confusion that multiple prey create when moving around, the dilution of risk when a predator attacks a larger flock, or the collective vigilance that emerges when animals are in groups. Regardless of the precise mechanism, for species that benefit from aggregation, predation rates may be density dependent, and at lower densities there may be a greater risk of predation (e.g., Sandin & Pacala 2005).

The relationships between time allocation and group size are referred to as **group size effects** (Bednekoff & Lima 1998; also see Chapter 7). Importantly, group size effects are not restricted to highly social species. Many species form transient foraging aggregations despite no long-lasting social bonds. Group size effects are typically studied by looking at the relationship between group size and time allocation to foraging and antipredator vigilance. The general assumption is that if group size provides safety, then we should expect to see that as group size increases, individuals allocate more time to foraging and less time to antipredator vigilance. In many species, we assume that foraging and vigilance are mutually incompatible; thus time allocated to vigilance cannot be allocated to other beneficial activities, such as foraging. However, this assumption does not hold in species (e.g., birds, lizards) in which the position of the eyes provides a wider visual coverage (Fernández-Juricic et al. 2004a) such that when an individual is head-down, it still can gather visual information from its lateral visual fields (Fernández-Juricic et al. 2005a, 2008; Tisdale & Fernández-Juricic 2009).

There are a few difficulties with quantifying group size effects. The first is that determining whether vigilance behavior is directed at predators or conspecifics is difficult and not always possible. There have been some attempts made with primates, by looking at gaze direction (Treves 2000), but most studies are unable to precisely identify the target of vigilance, particularly in species with laterally placed eyes. The second difficulty is that there are other reasons why animals might spend more time foraging in larger groups. For instance, if feeding competition increases with the density of conspecifics, then animals will forage more to get their fair share before food is depleted, rather than because of decreased risk. This is particularly a concern for species that engage in scramble competition (i.e., where individuals compete for a limited resource) on exploitable patches of food (Beauchamp 1998).

For species that benefit by grouping, social translocation or reintroduction may be essential. Some translocations and reintroductions fail because recently introduced individuals fail to establish a sustainable population as they are killed by predators. This creates an ethical issue—animals die because of our actions (Bekoff 2002); it also creates a practical one—the recovery may not work (Kleiman 1989). Doing anything to increase the survival of these animals is an important goal of much reintroduction and translocation research (Kleiman 1989). Thus, if animals are moved in social groups, predation rates may decline, and individuals may survive longer. The best evidence that social translocations may improve reintroduction success comes from studies of black-tailed prairie dogs (Shier & Owings 2007a,b). When intact social groups were moved, individuals survived longer because the likelihood of predation was reduced (Shier 2006; **Figure 10.2**).

Importantly, social translocation may avoid another problem: individually released animals returning to their natal areas. A major source of mortality of translocated carnivores is their leaving their translocation sites. The translocation of social groups in the 1995–1996 reintroduction of wolves to central Idaho and Yellowstone was intended to facilitate intact pack formation upon release into the wild. Translocation of packs is believed to con-

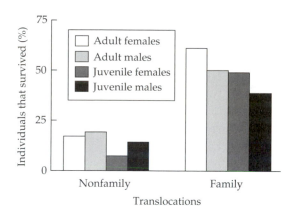

FIGURE 10.2   Black-tailed prairie dogs translocated in family groups had higher survival rates than those that were translocated in nonfamily groups. (After Shier 2006.)

tribute to successful territorial establishment and early breeding upon release from pens, as well as to curb wide-ranging movements that may have increased mortality risk, livestock conflict, and dispersal back to Canada after other releases (Bangs & Fritts 1996).

Conspecific Attraction

We discussed conspecific attraction in Chapter 5. Recall, the presence of other conspecifics may provide compelling evidence that a particular location is suitable. There is a growing body of evidence that animals use conspecifics as cues when assessing habitat suitability and making habitat settlement decisions (Stamps 1988). This phenomenon has been used by conservation biologists to help attract individuals to particular locations (Schlossberg & Ward 2004).

For instance, on the Fort Hood military base in Texas, conservation biologists were trying to recover the endangered black-capped vireo (*Vireo atricapilla*), a species that was negatively affected by brown-headed cowbirds (*Molothrus ater*) (Ward & Schlossberg 2004). Cowbirds are brood parasites that lay their eggs in other species' nests. Cowbird nest parasitism is responsible for the decline of a number of species, including the black-capped vireo. On the base, cowbirds were eliminated and wildlife managers wanted to attract black-capped vireos to nest in areas where cowbirds were controlled. To do so, they played black-capped vireo song from 4:00 to 10:30 A.M. during the nesting period. They found that in locations where song was played, more vireos nested and these nests were successful (**Figure 10.3**). Thus, by capitalizing on conspecific attraction, conservation biologists were able to locally recover a population.

However, social grouping (and conspecific attraction) may make a species more vulnerable to exploitation. Consider the large colonies of island-nesting seabirds and insular breeding grounds of marine mammals that made them ripe for exploitation by sailors upon their discovery. And

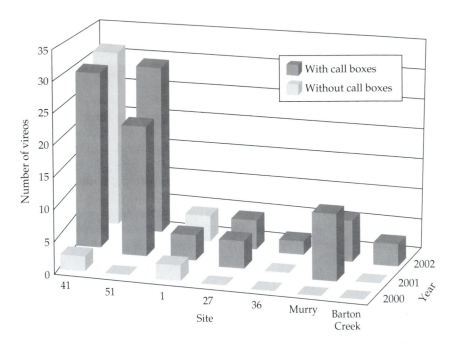

FIGURE 10.3   The number of vireos at different sites on the Fort Hood military base. In 2000, no conspecific calls were broadcast. In 2001 and 2002, calls were broadcast at some sites, and this lured birds into the nest boxes. (After Ward & Schlossberg 2004.)

consider herding ungulates, such as the plains buffalo (*Bison bison*), about which Richard Irving Dodge noted that in 1871, while he rode through a 25-mile-long herd, "the whole country seemed to be one mass of buffalo moving slowly northward" (Dodge 1877: 120). This species was hunted almost to extinction. Or consider the extinct passenger pigeon (*Ectopistes migratorius*), a species that lived in such large populations that "flocks in the migratory period partially obscured the sun from view" (Anonymous 1910). Habitat loss and easy hunting contributed to the decline of this once-abundant bird (Schorger 1955).

From the perspective of a hunter, hunting success may be higher on grouped than solitary individuals. This is because in some situations there is a positive density-dependent relationship between population size and predatory success (Sih 1984). Such a relationship may arise because individuals form search images for their prey, whereby hunting success increases with experience (which is correlated with population size), or because hunting success increases once a patch is located. For any given predator–prey system, it is an empirical question whether there is positive or negative density dependence, and the specific nature of this relationship may inform wildlife conservation.

Dispersal and Movement between Groups

In many species, dispersal is influenced by social structure or group size. Residents may disperse when local densities reach high levels and there are no breeding opportunities, but they may remain if there are opportunities within the high-density group. Such facultative dispersal increases the variation in the nature and types of social interactions found.

In some cases, wildlife managers wish to eliminate individuals that may have a transmissible disease. However, the outcome of such culling may be an influx of immigrants, and these immigrants may have to form social relationships anew, so such a strategy of killing residents may generate some drawbacks (Smith 2001). Localized killing of resident European badgers (*Meles meles*) to eliminate outbreaks of bovine tuberculosis (caused by *Mycobacterium bovis*) was found to be counterproductive. Badgers can be carriers of bovine TB, and for many years they were locally killed when infections were discovered. However, after these cullings, immigrants ranged widely and visited more setts (communal latrines) than residents before settling. This increased movement could have increased the rate of transmission of TB between badgers, and potentially from badgers to livestock (Woodroffe et al. 2005; Jenkins et al. 2007). Similar findings emerged from a study of brush-tailed possums (*Trichosurus vulpecula*) in New Zealand. Killing of residents led to increased movement of males in controlled areas, which potentially exposed more possums (and livestock) to TB (Ramsey et al. 2002). In such systems, reducing host movement, immunizing residents of the host species, and developing alternative methods to reduce livestock infection may ultimately prove more successful.

Phenotypic Plasticity in Social Structure

The social structure of many species is phenotypically plastic. Such intraspecific variation in social systems is thought to be adaptive (Lott 1991), but it raises the possibility that anthropogenic change can modify social structure. It could do so through at least two mechanisms: modification of the habitat or modification of the nature of the social relationships.

Habitat

Modification of the habitat is easier to envision, because many models of social evolution are based on the link between the distribution and abundance of resources and the social system. For instance, the classic Emlen & Oring (1977) model of mating systems is based on the idea that distribution of females is determined by the distribution of food, and the distribution of males is determined by the distribution of females. This is because female reproductive success is limited by food availability. Thus, females will be clumped if food is concentrated and will be dispersed if food is dispersed. Males need to eat, but reproductive success is limited by access to females.

If females are clumped and therefore defensible, polygyny may result. Thus, modifying the distribution of critical resources will change the distribution of females, and the mating system may vary.

We discussed this above, with respect to reproductively suppressed cooperative breeders limited by vacant territories or suitable nesting holes. But one could imagine manipulating other habitat features so as to manipulate breeding systems and the genetic consequences of these breeding systems. In their 2000 review, Maher & Lott identified a number of features that may influence spatial organization, including (but not limited to) food quantity, distribution, predictability, quality, renewal rate, type, density, and accessibility; resource distribution, quantity, predictability, and quality; population density; key habitat features; mate availability; sufficient space; refugia, spawing sites, or home sites; predation pressure; host nests; and energy availability. Well-designed experiments are needed to determine which manipulations are possible in a given system.

Social structure

To understand how human activity may affect group structure, a very brief introduction to social network analysis is required. **Social network analysis** is a tool that can be used to study the structure of groups. Social groups (and structure) emerge from interactions among individuals. In a network, individuals are nodes, and interactions between them form the links. By developing an association matrix of social interactions, one can plot the social network and calculate a number of network statistics about individuals as well as the overall group. These social network statistics formally describe attributes of sociality and, as such, provide a more comprehensive understanding of structure than simple measures like group size do (Wey et al. 2008).

One insight from social network analysis is that all individuals in a group may not be equivalent. A network approach to studying sociality suggests that certain individuals may be "key players" (Borgatti 2003). The removal of key players might have a disproportionate influence on social stability (**Figure 10.4**).

For instance, in pig-tailed macaques (*Macaca nemestrina*), adult males engage in third-party policing whereby they break up fights among females. By doing so, they have a stabilizing influence on the rate of agonistic interactions among females within a group. Interestingly, the importance of adult males is even greater than would be predicted by a network analysis. Flack et al. (2006) created experimental groups, observed social interactions, and then graphed the resulting networks. They then removed the data contributed by males and recalculated network structure (this served as a control). Compared with this control, the networks changed even more when males were experimentally removed from the social group.

In African elephants (*Loxodonta africana*), female matriarchs possess knowledge that helps increase the group's per capita reproductive success (McComb et al. 2001). These large individuals are often targeted by hunters,

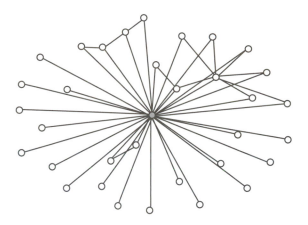

FIGURE 10.4 Illustration of a social network of a newly established group of social wasps (*Ropalidia marginata*). Circles represent individuals, and lines connect those pairs of individuals that interacted. You can see that the new queen (dark circle) is centrally located; her removal would fragment the group. (After Bhadra et al. 2009.)

and their removal has long-term deleterious consequences for young males that grow up without proper adult control/supervision and become real problems when they become adolescents (Bradshaw et al. 2005).

Position in a social network may influence demographically important events, such as dispersal. For instance, yearling yellow-bellied marmot females are more likely to remain in their natal group if they are socially well-integrated into it (**Figure 10.5**). Natal dispersal in yearling males, by contrast, is not influenced by social integration, because most yearling males disperse (Blumstein et al. 2009).

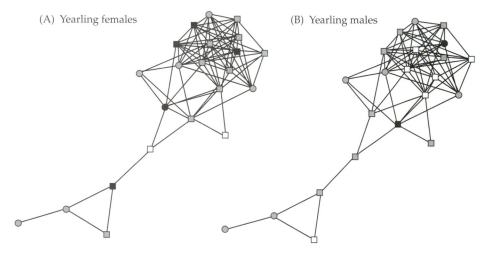

FIGURE 10.5 Affiliative social networks in yellow-bellied marmot females and males. Squares represent yearlings; circles, adults; white, dispersed; black, not dispersed; grey circles and squares represent other animals. (A) Female yearlings that dispersed (white squares) are peripheral to the group. (B) Male yearlings that dispersed (white squares) are not necessarily peripheral. (After Blumstein et al. 2009.)

Network analyses and analyses of the patterns of social interactions can be used to study the consequences of anthropogenic factors. For instance, social relationships can be influenced by human tourism, as was found in bottlenose dolphins (*Tursiops* spp.) by Lusseau (2003a). Specifically, Lusseau used a statistical technique called Markov-chain analysis to quantify how dolphin behavioral patterns were modified by ecotourists. He found that the presence of boats truncated dolphins' social interactions. Such interactions are essential for structuring the social group of this fission-fusion species (Lusseau et al. 2005). Lusseau (2003b) also used insights about the scale-free nature of the structure of dolphin social networks to predict that the loss of individuals would not fragment the cohesiveness of a group.

Applications

Mating systems, grouping patterns, and how individuals interact with each other influence survival and reproductive success. Thus, they are demographically important. Understanding the proximate mechanisms responsible for variation in such social behavior provides opportunities to manipulate it to achieve management outcomes. Many such manipulations can capitalize on the distribution of key resources, because resources have a profound influence on animals' spatial distribution. These resources include food, habitat, and conspecifics. Indeed, conspecific attraction capitalizes on the fact that conspecifics themselves are key resources. An understanding that social relationships can be influenced by anthropogenic factors (e.g., tourism) provides a window into possible management activities to maintain or to disrupt social structure. Because some management solutions require captive breeding, understanding social factors that influence reproductive success in captivity (e.g., infanticide and reproductive suppression) is important. In the next chapter, we go into more detail about reproduction and its relevance for conservation biology and wildlife management.

Further Reading

Several new books on social network analysis (Whitehead 2008; Croft et al. 2008) are excellent introductions into that literature and discuss programs that can be used to quantify social attributes. SOCPROG (http://myweb.dal.ca/hwhitehe/social.htm) is a freely available MATLAB-based software package that can be used to quantify the nature of social interactions. SOCPROG also comes in a compiled version that does not require a MATLAB license. UCINET (www.analytictech.com/ucinet) and NetDraw (www.analytictech.com/Netdraw/netdraw.htm) are widely used programs for the analysis and visualization of social networks. While NetDraw is free, UCINET has an inexpensive student license. KeyPlayer (www.analytictech.

com/keyplayer/keyplayer.htm) is a free program for identifying key players in a network. Reproductive skew is discussed in detail in Hager & Jones
(2009). Skew Calculator 2003 (www.eeb.ucla.edu/Faculty/Nonacs/
shareware.htm) is a freely available tool for calculating reproductive skew.

11 Demographic Consequences of Sexual Selection and Reproductive Behavior

There are population consequences of individual mating decisions (Smith et al. 2000). A variety of anthropogenic habitat modifications, including climate change, may directly or indirectly interfere with a species' natural reproductive behavior. They may do so by interfering with the mechanisms evolved to enable females to select their mates. This interference may result in a decline in reproductive success and thus a decrease in population growth rate. Interfering with evolved mate choice mechanisms may also create reproductive "mistakes" such as interspecies hybridization. Additionally, captive breeding that precedes reintroduction is notoriously difficult for some species, simply because individuals are not able to sample among prospective mates.

Knowledge of mate choice mechanisms may help us create more-natural social situations that encourage choice and increase mating success. And social and reproductive behavior may cause the violation of some assumptions of standard harvest models. For all of these reasons, we should understand the basics of reproductive behavior and sexual selection so that we can design effective mitigations, enhance reproduction when needed, and develop realistic harvest models.

In this chapter, we will introduce the concept of sexual selection and highlight its relevance for conservation biology and wildlife management. We will review the ways that animals chose their mates and then discuss how to determine how anthropogenic habitat modifications can interfere with reproduction. We will also discuss how mating systems can influence demography and discuss the consequences of some adaptive reproductive behavior for population harvesting models.

What Is Sexual Selection?

Darwin (1871) realized that in addition to natural selection, which could explain variation in survival, another process was needed to explain variation in mating success. He called this **sexual selection** and proposed that sexually selected characters evolve because of an advantage they give to members of one sex—usually males—in competing for or attracting mates.

Darwin suggested that traits such as horns, antlers, and aggressive displays evolved because they gave males an advantage in competing directly with other males for females; he called this **intrasexual selection**. This could come about in several ways: Males could dominate others and therefore have preferential access to females. Males could defend a territory and exclude competitors. Males could also sneak around in the shadows and not pay the costs of dominance or defense while surreptitiously cuckolding dominant males.

Darwin also identified **intersexual selection**, which is usually mate choice. Possible mechanisms of mate choice have been discerned in the years since Darwin. The examples that follow are from the perspective of the female as the choosy sex (but the logic applies to choices made by both sexes).

First, females may select males to obtain **direct nongenetic benefits**. For instance, if a male is parasite free, females will not get parasites while mating with him.

Second, there may be what is called a **Fisherian run-away process**. In this case, there must be a genetic covariance of the expression of a male trait (e.g., antler length) with the female preference for that trait. This creates a situation where parents produce sons with large antlers and daughters with a preference for large antlers. Given this genetic covariance, anything that increases male antler size will automatically select for the preference for male antler size.

Third, females may select males based on the genes that males carry (i.e., their "**good genes**"). Females may look for indicators of high genetic quality in potential mates. For instance, the ability to grow large antlers, or being brightly colored, may indicate that an individual can allocate sufficient resources to these sexually selected traits. Traits that reflect in an honest way the genetic condition of the male are sometimes called **condition-dependent traits**. Females mating with males with high-quality condition-dependent traits would have offspring with good genes, and selection is thought to act on females who are able to notice subtle differences in these traits and choose the good males.

Fourth, females may be duped. Sensory systems are used for multiple functions, such as finding food and choosing mates. Selection to enhance visual prey detection, for example, may have effects on female preferences for visual traits in males. Formally, genes that affect mate preferences have pleiotropic (i.e., multiple) effects on other sensory system functions. For instance, females may have evolved mechanisms to detect the movement and size of prey through feature detectors, which may also function in mate choice contexts. Male traits may evolve to take advantage of these female visual preferences. The idea of **sensory exploitation** is that male traits evolve to exploit these preexisting sensory biases or "hidden preferences."

It is very important that we understand what is and is not meant by *mate choice*. It is not necessary for females to be conscious of their mating decisions. In fact, it is not even necessary for females to make a decision at all. If a female's behavior in some way affects which males she mates with, this is considered female choice. Said this way, it is not much different from patch choice or even habitat choice. We are saying nothing about the cognitive abilities of an individual, just that an individual's behavior is influenced by what options it has.

Sexual selection thus favors traits that give the bearer an advantage in competing for or attracting mates. *Natural* selection favors traits that give the individuals carrying them a **fitness** advantage. Sexually selected traits are usually costly in terms of survival. But they work because fitness is a function of the number of offspring an individual produces per year multiplied by its breeding tenure. If some individuals have short but very productive lives, they can be as fit as individuals who breed less but live longer.

We shall focus mostly on intersexual selection (e.g., mate choice). While both males and females engage in mate-choice decisions, females, typically, are the choosier sex. Why? Males of many species often produce very small gametes (sperm) in a seemingly endless supply, while females of many species produce a much more limited number of larger eggs. This **anisogamy** may lead to sex differences in behavior (Bateman 1948; but see Tang-Martinez & Ryder 2005). In the case in which neither sex takes care of the young (e.g., in many insects and reptiles), females lay eggs and males fertilize them. For these systems, female fitness is limited by egg production, not mating success, while male fitness is limited by mating success, not sperm production. Thus, males are expected to compete for females, and females are expected to be choosy about their mates. Females who get only sperm from their mates may have nothing to gain by mating more than once, unless by mating multiple times they are engaging in cryptic choice. Indeed, mating more than once may even have negative effects on a female's fitness, because of time and energy wasted or extra exposure to sexually transmitted diseases. In these systems, male mating success should be more variable than female mating success (see the discussion of reproductive skew in Chapter 10).

How Do Animals Choose Their Mates?

Females rarely chose their mates randomly, and they often base their decisions on one or more identifiable male traits (Andersson 1994) and employ different rules to select among males. For instance, females may choose mates by using a relative or an absolute assessment rule. Relative assessment is inferred when a female's choice is based on a sample of available males; females using this mechanism should immediately change their preferences if offered a different selection of males (Zuk et al. 1990). While there is some empirical support for relative assessment mechanisms (e.g., Brown 1981), if females pay substantial assessment costs (e.g., time, energy, predation risk), they may have evolved absolute threshold assessment mechanisms to minimize assessment time (Wittenberger 1983). Such thresholds may be fixed or adjustable (Reid & Stamps 1997). When thresholds are fixed, females prefer particular expressions of male traits regardless of the distribution of those traits in the population (Zuk et al. 1990). When adjustable, thresholds may change, based on sampling of males in the population (Luttbeg 1996; Reid & Stamps 1997). Female pandas (*Ailuropoda melanoleuca*), for instance, spend more time investigating the scents of males that are deposited higher on trees (White et al. 2002). This is hypothesized to be a mechanism by which they identify the largest male.

Mechanisms of threshold assessment may be visualized with preference functions that plot the distribution of a male trait on the *x*-axis and the probability that a female will mate with a male with that trait expression (or a measurable correlate such as "interest," which may be estimated by the amount of time a female spends near a male) on the *y*-axis (**Figure 11.1**). If the plotted function were a line, the slope, *m*, would indicate the strength of the preference. Female preference curves may have different shapes. In one case, females may have asymptotic preferences—they prefer no males below a critical value and all males above another critical value. Between the critical values, the probability of mating increases with male trait expres-

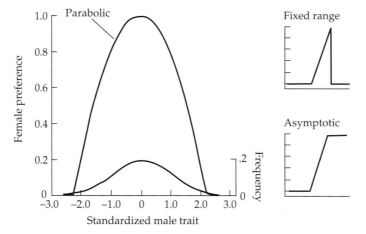

FIGURE 11.1   A parabolic female preference function (female preference axis) superimposed over a hypothetical male trait distribution (frequency axis). Female preference functions plot the probability of mating with a male exhibiting a particular trait expression (e.g., tail size). Females with parabolic preferences increase and then decrease their probability of mating with a male as male trait expression increases. Females with fixed-range preference functions (inset) mate only with males within a given range of trait expressions, and the probability of mating increases with male trait expression. Females with asymptotic preference functions (inset) mate with no males below a critical value and any male above another critical value. Between the critical values, the probability of mating increases with trait expression. (After Blumstein 1998.)

sion. In another case, females may have roughly parabolic preferences—
their probability of mating increases and then decreases as male trait expression increases. Parabolic-like preferences might be expected if female preferences covary with normally distributed male traits and each female has an
"ideal" mate (e.g., Jennions et al. 1995). Parabolic-like preference functions
would lead to stabilizing selection for male traits. Finally, although somewhat unlikely, females may have a fixed-range preference and mate only
with males within a given range of trait expressions. In this case, the probability of mating increases with male trait expression. While fixed-range preferences are unlikely, given that females seem to prefer exaggerated traits
(Ryan & Keddy-Hector 1992), they may potentially be found, depending on
the exact distribution of male traits (e.g., when preferences are asymptotic
but there are no highly desirable males present).

How do we study female preferences? Quantifying preference functions
is something best done with the control of captivity (see also Chapter 4). In
captivity, females might be given a "cafeteria" of males to chose from, and
either actual mating success or correlates of mating success, such as copula-
tion solicitation displays in birds or lordosis (a ritualized back-arching display) in mammals, would be scored. If male traits were systematically
manipulated (e.g., color changed, feather length modified, song or other
vocalizations experimentally broadcast) or if females were exposed to males
with different expressions of sexually selected traits, then it ultimately
would be possible to plot trait expression with probability of breeding or
attractiveness and thus create a preference function.

In nature, preference can be inferred correlatively. To do this, you
should identify most of the males in the population. Then, based on either
direct observations of mating or molecular reconstructions of parentage, it
should be possible to identify mates and thus determine whether males
with certain trait expressions are more likely to mate than others.

Female preferences are only half the story. Many of the traits that
females prefer vary according to male condition, and males in good condition may have longer tails and brighter colors and may vocalize more.
Andersson (1994: Table 6A) provides a comprehensive review of sexually
selected traits in both vertebrates and invertebrates. Of 232 reviewed studies on 186 species, 167 of the studies reported female choice for a trait. Many
of the listed traits used for mate choice are likely to be condition-dependent
traits. The expression of a condition-dependent trait may be influenced by
access to important resources, the presence of pathogens, or even the ability to properly display a trait (Andersson 1994). Thus, changes in resource
distribution, the introduction of an otherwise nonlethal pathogen, or habitat alterations may influence trait expression, and potentially mate choice,
and thus may have demographic consequences.

What happens if females have fixed-threshold preferences and a large
number of males are suddenly unable to acquire enough resources to grow
long tails, be bright, or vocalize a lot? Or what happens if females have a
flexible threshold that requires some sampling but they are unable to

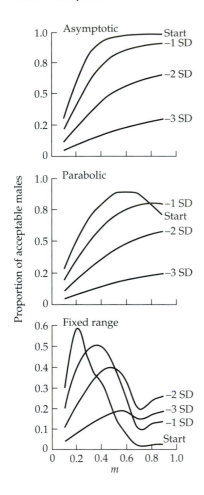

FIGURE 11.2 Proportion of acceptable males as a function of the "steepness" of the preference function, m (calculated by regressing male trait distribution against female mating probability), for each of the three different preference functions, after shifting the distribution of male traits 1 SD, 2 SD, and 3 SD "downward." With asymptotic and parabolic preferences, there is almost always a substantial decline in the proportion of acceptable males as the distribution of the trait that the females are choosing gets smaller. (Modified from Blumstein 1998.)

sample? A long winter, the sudden introduction of new parasites or pathogens, human impacts, or natural disasters may all reduce food availability or condition prior to a mating season, and they may result in uniform downward shift in the expression of condition-dependent traits or may even increase the cost of sampling.

A simple model (Blumstein 1998) explored the consequences of this problem and found that for those species in which females have fixed-threshold preferences for male condition-dependent traits, a sudden change in the phenotypic distribution of those male traits may cause a decline in the number of acceptable males and a concomitant decline in the effective population size, N_e (**Box 11.1**).

Specifically, shifting the mean value of male trait expression so that it was up to 3 standard deviations smaller than the original mean value led to a substantial decline in the proportion of acceptable males (**Figure 11.2**). And as the proportion of acceptable males declined, the N_e/N ratio declined (**Figure 11.3**). This means that the population size that you have is really much smaller than you think.

The effects of small population sizes are well known (Frankham et al. 2002). As N_e decreases, genetic variation is lost, the likelihood of inbreeding and its potentially deleterious effects increases, and the likelihood of random demographic effects that lead to local extinction increases. Reviews of N_e/N ratios suggest that N_e in wildlife populations may already be very

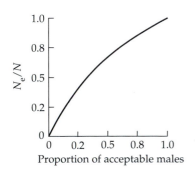

FIGURE 11.3 The relationship between the proportion of acceptable males and N_e/N in a population that is not age structured. As fewer males are acceptable, N_e/N decreases more rapidly. (After Blumstein 1998.)

BOX 11.1 Effective Population Size, N_e

The number of individuals in a population, N, is a first approximation of endangerment. However, other factors influence the likelihood that a population will go extinct over time. For instance, variation in the number of breeding individuals, variation in breeding success, the ratio of breeding males to females, and other factors influence the maintenance of genetic variation in a population (Falconer 1989). Genetic variation influences long-term sustainability because genetic variation is required to combat any negative effects of inbreeding and to allow evolutionary adaptation to an ever-changing environment.

To compensate for the inadequacy of N alone in predicting the likelihood that a population will persist over time, population geneticists have developed the concept of the **effective population size**, N_e which is a better predictor (Gilpin & Soulé 1986). N_e is an estimate of the theoretical number of breeding individuals, assuming they behave in an ideal way.

N_e models an ideal population with the following properties: the population is split into subpopulations and there is no migration between subpopulations, generations do not overlap, the number of breeding individuals is the same for all generations and subpopulations, mating is at random and includes a random amount of self-fertilization, there is no selection, and mutation is assumed to be unimportant (Falconer 1989).

Even with these simplifying assumptions that may not normally be met in real populations, it is possible to calculate N_e in real populations. N_e gives us an indication of what the breeding population size really is. N_e is inevitably smaller than N, and the smaller this ratio, the less the genetic diversity in the population.

As N_e declines, population viability is negatively affected if homozygosity is increased and if the number of nonselected alleles is decreased. The loss of variation is compounded by an increase in linkage disequilibrium (nonindependent assortment of alleles), which reduces the frequency of novel gene combinations. N_e is influenced by factors that halt the passing of genes to the next generation. Falconer (1989) identified six of these factors: (1) exclusion of closely related matings, (2) skewed sex ratios, (3) unequal generation size, (4) unequal family size, (5) inbreeding, and (6) overlapping generations.

A standard equation to calculate the effective population size for a population that is not age structured is (Wright 1938):

$$N_e = \frac{1}{\dfrac{1}{4N_m} + \dfrac{1}{4N_f}}$$

where N_e = the effective population size, N_m = the number of breeding males in the population, and N_f = the number of breeding females in the population.

Thus, calculations of N_e explicitly acknowledge that not all individuals breed (e.g., Parker & Waite 1997). The implicit assumption is that most breeding-age females breed but only a fraction of breeding-age males breed.

The relevance of this concept is that mating systems directly affect the genetic resilience of a population (see text).

small. Thus, anything that can cause a further decline in N_e is worthy of attention, particularly if the effects of a declining N_e are not predictable (Frankham et al. 2002).

We know that the expression of a variety of male traits may be influenced by the male's condition (Andersson 1994). Unfortunately, we know relatively little about the heritability of female preferences (but see Andersson 1994), nor do we know whether females who appear to have fixed thresholds can, or will, change their preferences when faced with an array of formerly unacceptable males (Forsgren 1992). Thornhill (1984) provides evidence of some preference flexibility on an ecological timescale for a species with threshold assessment mechanisms, and Reid & Stamps (1997) illustrate an example of an adjustable threshold, but more data are required to generalize about responsiveness.

However, especially for species with brief fertile periods, it is possible that females do not have enough time to respond to a sudden decline in male "quality." If so, N_e will drop in three ways: (1) there will be fewer acceptable males, (2) females may be unable to find suitable mates while they are fertile, and (3) some females may choose not to breed. Actually, failure to breed in a given year is not unheard of in natural populations (e.g., Boag & Grant 1981). Moreover, female condition and fertility are also likely to covary with the same environmental resources that influence male condition; females in poor condition may need to spend more time foraging and less time assessing male quality. If females are unable to breed, the effect of a decline in acceptable males will be magnified, and N_e will decline even more.

Perhaps the above scenario is an argument against a species' having a fixed-threshold assessment mechanism if females base mate choice on variation in condition-dependent traits. In other words, we might not expect the evolution of a fixed-threshold assessment mechanism where failing to reproduce is a common occurrence. Nevertheless, in lieu of specific information about mechanisms of female choice, we should consider fixed thresholds a possibility because of their profound potential to influence breeding success.

There are important ramifications for the management of threatened or endangered species if female mating preferences can contribute to a rapid decline in N_e. First, more must be known about female assessment mechanisms. Behavioral ecologists can make an important contribution to conservation biology by studying the mechanisms of female mate choice in more detail. With this knowledge, traits may be manipulated (e.g., tail lengthened, color changed) to increase mating probability. Second, care should be taken to avoid situations where condition-dependent male traits will not be fully expressed during breeding seasons. For instance, translocations and reintroductions (moving animals from one location to another to recover a locally extinct population; Kleiman 1989; Griffith et al. 1989) or the transfer of reproductive individuals between zoos for captive-breeding programs may be timed to avoid stressing males in ways that may influence their physical condition.

Genetic Consequences of Mating Systems

Species (and sometimes individuals within a species) vary with respect to their mating systems (see Chapter 10). Patterns of reproduction in a group will influence the genetic properties of the subpopulation and population. Traditionally, population geneticists focused on three types of heterozygosity: variation within individuals, variation among individuals in the same subpopulation, and total population variation (Sugg et al. 1996). Sewall Wright (1969, 1978) developed fixation indices (F_{IS} within subpopulations, F_{ST} among subpopulations, and F_{IT} within the entire population) that calculate an observed value of genetic variation compared with what would be expected in an ideal population (e.g., with random mating and no mutation or selection). These fixation indices can be used to estimate the degree of inbreeding and therefore the rate at which genetic variation is lost.

Species with social structure, however, violate a fundamental assumption of such models, namely that animals mate randomly within a subpopulation (Sugg et al. 1996; Dobson et al. 1998). Does this make a difference in predicting the rate at which genetic variation is lost? Sugg et al. (1996), Dobson & Zinner (2003), and Dobson (2007) reviewed a series of studies and concluded yes. The key is that kinship relationships (coancestry), which develop from mating tactics and sex-specific dispersal strategies, develop more quickly than inbreeding in social groups. Historically, kinship and inbreeding were the only mechanisms proposed to account for genetic similarity. Yet, the formal models developed to study gene dynamics describe how genetic variation changes as a function of social structure. And it is this genetic structure that it is essential to quantify if one is to properly calculate effective population sizes.

The argument is based on a formal and complex model called the breeding-group model (Chesser 1991a,b; Sugg and Chesser 1994). Sugg et al. (1996) evaluated the breeding-group model by applying it to black-tailed prairie dogs (*Cynomys ludovicianus*). These social rodents meet a key assumption of the model: populations are subdivided into groups composed of kin. They found that in prairie dogs, the rate of female dispersal influenced the rate at which genetic variation was lost (Dobson et al. 1997, 1998, 2004). As female dispersal rates increased, heterozygosity was lost faster. This finding has a somewhat counterintuitive implication for a commonly used population recovery technique: translocation. If females were moved from their original colonies to found new colonies, genetic variation would be more quickly lost. Without going into the details of the model here, their results suggest that it is essential to understand the effect of sociality on gene dynamics when designing management strategies to preserve genetic diversity (see also Chesser et al. 1996). Those interested should read these papers closely.

Mate Choice Copying: A Particularly Important Mate Choice Mechanism for Conservation Behavior

When you make a major purchase (e.g., you buy a car), do you read everything written about all models, or do you talk with trusted friends about their opinions? Talking with trusted friends is a good way to reduce assessment costs (the time and energy you spend searching for a car). Females of a variety of birds (Gibson & Höglund 1992) and fish (e.g., Dugatkin 1992; Schulpp et al. 1994) do the same thing: rather than independently assessing all possible males, they copy the mate choice decisions of others (**Figure 11.4**; Gibson & Höglund 1992).

Mate choice copying reduces the costs of independent assessment. This is important if assessment is risky or time-consuming. For instance, if by searching out and investigating a large number of males, females expose

FIGURE 11.4   In birds that use leks (traditional mating sites), such as the greater sage grouse (*Centrocercus urophasianus*) illustrated here, females may copy the mate choice decisions of other females, which may reduce female assessment costs. Such copying leads to a decline in the effective population size because only a subset of males reproduce. (Photo courtesy of Neil Losin.)

themselves to a greater risk of predation, it might behoove them to copy the decisions of another female. In many circumstances, time is of the essence. In such cases, extensive time spent making a decision might make it so that a female's young are born too late in the season to have a high probability of survival. For inexperienced, juvenile females, who might not be as savvy about selecting mates as older, more experienced females, copying the mate choice decisions of an elder might be beneficial.

Mate choice copying is an evolved mechanism that may lead to different types of preference functions. Specifically, preference functions may not be stable, and a female's preference may depend upon her experience. Thus, if we see inconsistent choices, we may have a hint that females are copying the behavior of others. Another hint that mate choice copying may be occurring is seeing very biased male reproductive success (as would be predicted if females were copying the decisions of other females).

What are the implications for conservation behavior? One big implication is related to captive breeding. Whenever there are hints that females use other females as cues when making mating decisions, it is likely that captive-breeding success will increase if animals are allowed to select their mates socially. It is possible, yet unstudied, that species that rely on mate choice copying also have socially facilitated fertility. In such cases, we might have to house animals socially to maximize fertility. Finally, copying is a mechanism that may reduce genetic variation, and identifying factors that reduce variation is important for genetic management.

Anthropogenic Changes Interfere with Reproduction

Female preferences, which may create or maintain species diversity, can be affected by human activities. For instance, the number of species of African cichlid fish living in Lake Victoria has been declining for some years. Seehausen et al. (1997) provide compelling evidence that the loss of diversity is related to increased turbidity, a consequence of farming practices, which decreases female visual acuity and also prevents females from expressing preferences for certain colors (**Figure 11.5**). In the cichlids' case, females hybridize with other species, leading to a loss of species diversity. If hybrids were not fertile, it would lead to an immediate reduction in the number of fish as well. Neither the importance of direct impacts by humans on cichlids nor less direct effects via mate choice should be ignored. Sexually dimorphic species may have sexually selected traits that may reduce a species' competitive ability or make it particularly vulnerable to human impacts that modify the expression of those traits.

Fisher et al. (2006) found another effect of human impact. Female swordtails (*Xiphophorus birchmanni*) living in clean water have a strong preference for mating with conspecifics. However, when females were tested in water polluted with sewage or agricultural runoff, they lost their discrimination for conspecifics over heterospecific *X. malinche*. The authors inferred that this resulted from the humic acid contained in human and agricultural

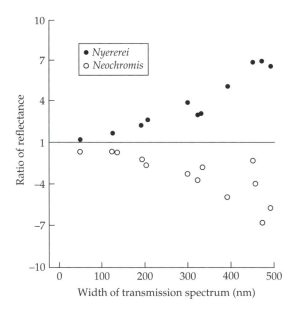

FIGURE 11.5   The effect of turbidity on the distinctiveness of two species of Lake Victoria cichlids. The *x*-axis plots the width of the light transmission spectrum in the lake water. At sites where the lake was more eutrophic, less light was transmitted, so the transmission spectrum was narrower. The *y*-axis plots the distinctiveness of a species with red nuptial coloration (top) and a species with blue nuptial coloration (bottom). When the ratio of reflectance is 1.0, there is no difference in the reflectance of blue and red. Thus, as sites became more eutrophic, it became more difficult for fish to discriminate among species. The black circles plot the brightest "red" males, and open circles plot the brightest "blue" males along a eutrophication gradient. As males became less distinctive, hybridization ensued. (After Seehausen et al. 1997.)

waste and conducted an experiment to demonstrate that humic acid could interfere with chemical communication systems and lead to hybridization. Chemical communication is a prevalent sensory modality, not only in fish, but also in many mammals, including highly social ones (e.g., elephants; Schulte et al. 2007). Understanding the mechanisms of mate choice can help target remedial work to reduce its impact.

How Mating Systems May Influence Demography

Mating systems affect N_e by determining which genes are passed on to the next generation (Parker & Waite 1997; Greene et al. 1998). This happens both directly, because of limited breeding, and indirectly, because in systems where there is extreme skew, animals are competing for breeding opportunities and they experience higher mortality associated with this competition. Mating systems can also have a large effect on the way population size responds to natural and human-induced habitat changes, hunting, and so forth (Parker & Waite 1997; Greene et. al. 1998; Whitman et al. 2004; Maldonado-Chaparro & Blumstein 2008), and they limit the distribution of species (Höglund 1996). Wildlife managers need to incorporate knowledge of mating systems into management plans (Greene et al. 1998; Caro et al. 2009).

Using a modification of the equation in Box 11.1, Parker & Waite (1997) illustrated how mating systems influence N_e. Assuming an N of 100 and a reproductive failure rate of 50%, a species displaying promiscuous mating (every male mates with every female) has an N_e of 67. Under the same conditions, a monogamous species (a single male mates with a single female) has an N_e of 50, and an extremely polygynous species (males mate with multiple females, and not all males mate) has an N_e of only 19, while an extremely polyandrous species (females mate with multiple males, and not all females mate) has an N_e of only 9. Information about mating systems can be used by managers to understand the differential effects of habitat destruction or harvesting on N_e of different populations.

For example, Greene et al. (1998) modeled the effect of mating system and several other variables (harem size, infanticide, and reproductive suppression) on how a population would respond to three different types of hunting: trophy hunting of adult males only, hunting of adults of both sexes, and indiscriminate subsistence hunting. All forms of hunting are more detrimental to monogamous species than polygynous species: populations of monogamous species grew at slower rates when hunted. Hunting of only males had the greatest effect on infanticidal species because it increased the chance of sexually selected infanticide (see below) and thus reduced population growth rate. Reproductively suppressed species (species where not all adults reproduce) were most impacted by hunting of adults, because a population with reproductive suppression contains many nonbreeders. This mimics the situation of a series of monogamous groups, and thus the elimination of adults had a large effect on the population growth rate.

Mating systems may vary within and between species (Lott 1991). Such variation may be adaptive in the sense that intraspecific variation is driven by the same ecological factors that influence interspecific variation. In cases where there is intraspecific variation, it may be difficult to always predict the relationship between N and N_e for a given species. Thus, harvesting species with variable mating systems may require that hunting laws be tailored to each population, as well as varied over time.

Sexually Selected Infanticide by Males: An Evolved Strategy to Increase Male Reproductive Success

As we talked about in Chapter 1, in many species, when a new dominant male joins a social group or acquires a previous male's territory or status, the new male kills offspring sired by the previous male, a phenomenon known as **sexually selected infanticide** (Hrdy 1979). While it was initially thought aberrant and maladaptive, research in the past three decades has shown that for many species, sexually selected infanticide is an adaptive reproductive strategy because females, which quickly go into estrus after their offspring are killed, sire young that are protected by the new male (Hausfater & Hrdy 1984). Sexually selected infanticide influences N_e directly, by reducing juvenile survival and thus N, as well as indirectly, by reducing recruitment and therefore population growth rate, r.

Infanticide occurs in a wide variety of taxa and can cause as much juvenile mortality as predation does. For instance, 50% of young golden marmots, *Marmota caudata aurea*, that emerge aboveground one year survive to the next year (Blumstein 1997). At least 22% of this first-year mortality appears to be caused by infanticidal male marmots, while predation is responsible for another 22%. New male immigrants apparently kill unrelated pups soon after they visit or join a social group. This pattern is not restricted to marmots. While infanticide is best documented in rodents (Ebensperger & Blumstein 2007) and primates (van Schaik & Janson 2000), it is also seen in a variety of taxa, including canids, felids, birds, and fish (van Schaik & Janson 2000).

Infanticide has important implications for management. Traditionally, hunting males was not thought to decrease N_e because there are often unmated males that quickly replace missing males. However, in Sweden, where brown bears (*Ursus arctos*) are hunted, killing one adult male may lead to the deaths of nonhunted juveniles because other males may then kill the offspring sired by the previous male, with a resultant reduction in recruitment. Swenson et al. (1997) calculated that the demographic effect of killing a single adult male was equivalent to killing 0.5–1.0 adult female. Similarly the removal of harem-holding male lions (*Panthera leo*) at low harem sizes allows unrelated males to take over the harems and kill offspring (Greene et al. 1998). It is likely that sexually selected infanticide may have similar effects on other species.

In muroid and sciurid rodents, Blumstein (2000) found that infanticide by males was an ancestral trait that was lost in some taxa. Wildlife biologists using a phylogenetic approach (see Chapter 3) to understand the behavior of a focal species would thus hypothesize that in some taxa, it might be prudent to assume that sexually selected infanticide may be present, and then manage accordingly. For instance, with no other knowledge about a muroid or sciurid rodent, we should not expose females with young to novel males. This means that the timing of translocations or reintroductions should be worked out so that young of the year are able to defend themselves before new animals are introduced to an area.

Management sometimes requires that individuals be moved. Movements, whether translocations or reintroductions, should be carefully evaluated with respect to whether or not a species is likely to exhibit sexually selected infanticide. In some cases, moving females may be less risky than moving males who may, upon finding themselves in a novel social situation, kill unrelated young. Another situation where knowledge of infanticidal tendencies could be important is when habitat modifications change movement patterns. For instance, designing a corridor may have unexpectedly large impacts on population growth rate if new males move through the corridor killing unrelated young that they encounter or if increased movement allows greater contact between infanticidal males and family groups. Managers must consider whether or not a species of interest is infanticidal, especially when considering harvesting laws.

Applications

Mate choice can be manipulated for conservation purposes. Fisher et al. (2003) developed an ingenious idea based on a fundamental understanding of chemical communication and sexual selection. They assumed that females were looking to mate with high-quality males. High-quality males should be able to repeatedly scent mark an area. Thus, they assumed that females would prefer to mate with males whose scents were most frequently encountered. They tested this idea with captive pygmy lorises (*Nycticebus pygmaeus*), a nocturnal species of conservation concern. For several weeks prior to a mate choice trial, females were exposed to the urine of a particular male. The researchers found that females were more likely to investigate a "familiar" male than an unfamiliar male. Fisher et al. (2003) suggest that these sorts of manipulations can help wildlife biologists pair individuals in such a way as to increase N_e in captive breeding and potentially in nature.

As discussed in this chapter, such manipulations need not be restricted to mate choice (whether in captivity or in the wild). A more fundamental understanding of how pollution and other anthropogenic factors interfere with reproduction can create concrete targets for pollution and disturbance control if the objective is to avoid hybridization and enhance population viability. In addition, the proportion of hybrid individuals may be used as a bioassay of functional interference with reproduction. Finally, reproductive success is influenced by social factors. Thus, both habitat modifications (also discussed in Chapter 10) and changes in harvesting regimes may have profound implications on fitness and population viability.

Further Reading

This chapter was expanded and modified from Blumstein (1998) and Anthony & Blumstein (2000), both of which are good sources to consult for more information about the behavioral side of these topics. Frankham et al. (2002) is an outstanding text on conservation genetics.

12 Using Behavior to Set Aside Areas for Wildlife Protection

The Problem

A central concern in conservation biology is minimizing the effects of human disturbance (e.g., habitat fragmentation, urban sprawl, recreational activities) on species of conservation concern and on animal communities. Recreational activities may have deleterious impacts on biodiversity, particularly in developing countries where ecotourists go to experience biodiversity. The rate of human visitation to the world's biodiversity hotspots is expected to double by 2020 (Christ et al. 2003). The effects of recreational activities on birds have been widely documented and include changes at the individual, population, and community levels. These changes may include such things as increased antipredator vigilance (Burger 1994), reduced foraging time (Fernández-Juricic & Tellería 2000; Fernández-Juricic et al. 2003), increase in stress hormone levels (Müllner et al. 2004), reduced mating activity (Gutzwiller et al. 1994; de la Torre et al. 2000), reduced parental investment (Piatt et al. 1990; Lord et al. 1997; Ghalambor & Martin 2000), increased nest guarding (Komdeur & Kats 1999), reduced overall survival and reproductive success at the population level (Morris & Davidson 2000), reduced nesting success (Beale & Monaghan 2004), and changes in species composition at the community level (Fernández-Juricic 2002; Crowder et al. 1997; Morris & Davidson 2000).

One of the many management strategies to reduce these deleterious effects is to limit the access or amount of human visitation to a protected area. There are two major questions when it comes to constraining human disturbance. The first is: How much area should human activity be banned from, and for what duration? The second is, assuming that recreation is controlled spatially and temporally: What is the maximum visitation level that a species can tolerate without major consequences to survival, reproductive success, and persistence at the local scale? These questions have been the subject of considerable research efforts, mostly from the perspective of documenting patterns of change in individuals, populations, and communities and their correlation with various levels of recreational activities. However, relatively less effort has been devoted to identifying the behavioral mechanisms that drive these patterns. Understanding mechanisms is important because they can provide a way of predicting the responses of wildlife to different zoning scenarios.

A Theoretical Framework

Frid & Dill (2002) proposed a framework for understanding (and predicting) the responses of wildlife to recreationists. This **risk-disturbance hypothesis** assumes that animals seek a balance between avoiding disturbance and pursuing activities that may increase fitness, such as foraging, mating, and parental care (Frid & Dill 2002). Consequently, animals are expected to optimize their behaviors to minimize risk but still maintain access to resources (e.g., tolerate closer human approaches when closer to a refuge). The risk-disturbance hypothesis predicts that animals will respond to short-term changes in spatial and temporal patterns of disturbance according to their perception of risk. Riskier situations are expected to translate into greater responses at different levels of ecological organization, including the individual level (behavior, physiology), the population level (changes in abundance, recruitment), and the community level (changes in species richness, composition, extinction, colonization). One of the assumptions of the risk-disturbance hypothesis is that animals' reactions to humans are similar to their reactions to predators—they become alert and flee eventually if the costs of a human approach are higher than the benefits of staying in the patch (Frid & Dill 2002). This assumption has some empirical support (Fernández-Juricic & Tellería 2000; Møller et al. 2008).

Reactions to humans can have different consequences if different species respond differently. Lima & Dill (1990) and Caro (2005) described the different phases involved in an encounter between a predator and a prey. These phases can be summarized as *prey first detecting a predator* and then *reacting to it*. A similar idea can be applied to the detection and escape from recreationists. From a habitat-use perspective, when animals detect (and become alert to) humans, activities that animals gain fitness through (e.g., foraging, mating) may be disrupted so they can track the source of disturbance. If these interruptions are frequent, then some patch use (e.g., Fernández-Juricic & Tellería 2000) and physiological parameters (e.g., stress hormones; Tarlow & Blumstein 2007; Busch & Hayward 2009) may be affected.

However, this may not necessarily generate a species-level impact (i.e., a species negatively affected by human disturbance across its whole distributional range). The response may only be local and restricted to a single population with high levels of recreationists. And a local, population-level response may not even be detectable if animals do not have alternative habitats to use, as they may tolerate the high levels of disturbance and remain on a particular patch (Gill et al. 2001). Such animals may **habituate** (i.e., decline in responsiveness following repeated stimulation) or alternatively **sensitize** (i.e., increase in responsiveness following repeated stimulation). If they become more reactive to humans, they could experience reduced fitness.

After predator/human detection, and depending on the direction and speed of the threat, animals may flee to an adjacent refuge until the disturbance goes away, or they may leave the patch for a different one within the same geographical area. The latter scenario can have consequences for the amount of suitable habitat present if, because of fear, animals refuse to use patches that would otherwise be suitable.

The risk-disturbance hypothesis provides a strong conceptual framework for understanding spatially explicit human–wildlife interactions. The **resource-use–disturbance trade-off hypothesis**, a special case of the risk-disturbance hypothesis, predicts patch use based on the relationship between the frequency of resource use by animals and the frequency of human visitation (Fernández-Juricic 2000, 2001; Fernández-Juricic et al. 2003). This hypothesis predicts that relatively low pedestrian rates may not generate negative impacts on wildlife, as long as the rate of disturbance (e.g., number of recreationists per unit time per unit area) allows animals to cope with their breeding and feeding requirements in the intervals between consecutive visits. However, when pedestrian rates increase, the reduced lengths of intervals between visits may decrease the temporal and spatial availability of resources for animals (**Figure 12.1**). Such disturbance could potentially reduce the suitability and carrying capacity of disturbed areas, with consequent reductions in density (Fernández-Juricic 2000). This is a mechanism that explains human–wildlife interactions at very local scales (e.g., the patch level), and it mostly affects individuals. For this mechanism to affect wildlife at the population level, disturbance events have to be widespread in time and space in an area with suitable habitat. For instance, the Iberian frog (*Rana iberica*) reduces its stream bank use as the intervals between consecutive visits by recreationists become shorter (**Figure 12.2**), which in turn leads to a decrease in population density on stream banks closer to recreational areas (Rodríguez-Prieto & Fernández-Juricic 2005).

The resource-use–disturbance trade-off hypothesis assumes that the effects of human disturbance are only apparent when pedestrians are within an individual's detection window (Fernández-Juricic et al. 2001a; Fernández-Juricic & Schroeder 2003), which triggers vigilance and potential escape. A **detection window** is an area around the animal within which the chances of detecting a source of disturbance (visually, acoustically, chemically, etc.) are higher than outside of it. This idea is important because the sizes and shapes of detection windows differ among species, probably based on a species' main sensory modality (Blumstein et al. 2005). The implication is that similar levels of recreational activities could have different effects depending upon, among other things, the ability of different species to detect humans.

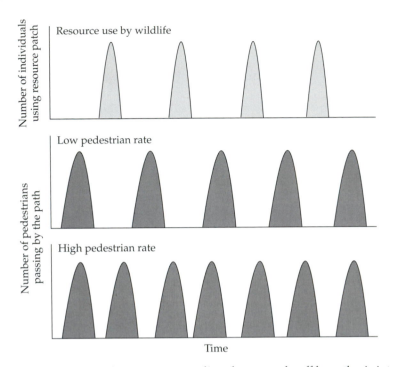

FIGURE 12.1   The resource-use–disturbance trade-off hypothesis intends to predict the responses of wildlife to changes in the local frequency of disturbance. Imagine a scenario in which wildlife use the resources of a given patch a number of times during the day (light curves), and recreationists walk around that patch a few times during the day (dark curves). At low pedestrian rates, local (patch) exploitation of a given resource by animals can be accomplished in the intervals between consecutive visits by pedestrians, allowing coexistence between wildlife and recreationists. However, at high pedestrian rates, animals showing antipredator behavior to humans may refrain from exploiting a given patch because of the lack of disturbance-free time. If this local situation is spatially widespread (recreationists allowed full access to protected areas), foraging and breeding opportunities could be substantially reduced.

Under which applied contexts are these hypotheses useful? Theory assumes that the behavioral interactions between humans and wildlife will influence individual behavior, potentially affecting populations if disturbance is widespread. Therefore, these hypotheses can help us understand how to manage for coexistence between wildlife and humans in a given area. For instance, we can predict the minimum distances at which humans can approach wildlife before disturbing them so that we can develop guidelines that facilitate the presence of a species of interest and the potential for bird-watchers to see/hear that very species. This can enhance the local educational and economic value of a protected or natural area. In addition, by considering the sizes and shapes of detection windows of different species, we may be able to predict responses of species with similar food and nest-

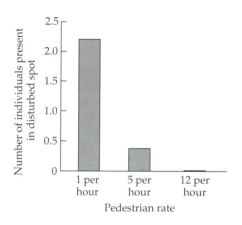

FIGURE 12.2 Number of Iberian frogs observed on a stream bank under different levels of recreational activity. As the interval between recreationists passing by decreases (increase in pedestrian rate), the number of detectable frogs decreases. (After Rodríguez-Prieto & Fernández-Juricic 2005.)

ing requirements, which may facilitate managing groups of species rather than single species within protected areas (Blumstein et al. 2005). Finally, this approach enables us to have a critical look at how to improve some conservation tools, such as buffer areas.

Applications

Now, imagine you are in charge of managing an area that is also used for recreational bird-watching, hiking, horseback riding, and mountain biking. A new survey discovers a threatened species in this area. You are asked to define "no-go zones" to protect this species but still allow for recreational activities. How might you estimate the amount of land to be set aside to minimize the effects of human disturbance on this threatened species? The answer is relevant for both conservation biologists and landowners. If we underestimate the amount of area to be protected, it may lead to increased human impacts on wildlife. If we overestimate the amount of area to be protected, it may result in high economic losses due to unnecessary litigation over the allocation of space to different human activities. This is because those who will most likely be affected will be developers, large industries, and those companies that make use of natural resources.

Land managers created the concept of buffer areas to establish restrictions on the human use of some areas. However, the terminology describing buffers for wildlife protection has not been consistent in the literature, which sometimes leads to some confusion. The buffer area concept has been applied both at the landscape scale (hereafter, *landscape buffer area*) and at the patch scale (hereafter, *patch buffer area*). A **landscape buffer area** (also known as a buffer zone) is an area that surrounds critical habitat for a population or a community so that it can serve as an environmental "cushion" to minimize external direct or indirect human disturbances (Reid & Miller 1989). Landscape buffer areas surround core sectors of protected areas, providing additional habitat for species protection (Shafer 1999a). They may also be used for some specific activities not allowed in core sectors, such as recreational activities and controlled extraction of natural resources (Sayer 1991; Shafer 1999b). On the other hand, a **patch buffer area** (also known as a setback zone) denotes a minimum area of critical habitat for an individual or a group of individuals concentrated in space (breeding colony, roost; Vos et al. 1985; Fox & Madsen 1997; Fernández-Juricic et al. 2005c) in which encroachment would trigger negative effects on wildlife.

Estimations of landscape and patch buffer areas have involved a combination of expert judgment and the application of some simple geometric calculations. However, the dimensions of a landscape buffer area may vary depending upon the management goal (e.g., maintaining natural fire regimes or species diversity), and those of a patch buffer area may vary according to the spatial distribution of the target species (e.g., colonial, territorial). We sorely need further research on improving the estimation of buffer areas at the landscape and patch scales, taking into consideration biological parameters that influence the spatial and temporal patterns of microhabitat selection as well as the detection of and responses to recreationists by species with different life histories and sensory systems (Fernández-Juricic et al. 2005c). We now present a few examples of how to estimate buffer areas.

One way of estimating a landscape buffer area is by measuring the distance from a source of disturbance (e.g., visual, acoustic) to the point at which the density of a species or the species richness increases substantially, because this distance suggests that the disturbance has lessened in strength (Reijnen et al. 1995; Miller et al. 1998). Usually, landscape buffer areas are calculated from a threshold distance, a minimum width of habitat required for protection. For example, Palomino & Carrascal (2007) found that the abundance of birds in the highly urbanized landscape around Madrid, Spain, varied in relation to the cover of deciduous trees and the distance to the roads (potential sources of acoustic disturbance; **Figure 12.3**). Based on these relationships, they established landscape buffer distances of approximately 300–400 meters from urban areas and roads to protect bird diversity.

Conceptually, the estimation of patch buffer areas is related to the risk-disturbance hypothesis (Frid & Dill 2002) because it assumes that individuals perceive humans (or human-related activities) as threats and respond to them by becoming alert and eventually fleeing. Patch buffer areas are based on estimates of the distance at which animals become fearful of humans through a two-tiered approach (Knight & Skagen 1988; Knight & Temple 1995; Richardson & Miller 1997). First, the **minimum approaching distance** is calculated, that is, the minimum horizontal distance of nonintrusion by humans that would preclude disturbance (Rodgers & Smith 1995; Rodgers & Schwikert 2002). Second, individual buffer areas are generally estimated using the minimum approaching distance as the radius of a circle (Fox & Madsen 1997). Interestingly, patch buffer areas have been found to vary between species and, specifically, to increase with a species' body size. For instance, buffer areas of mountain woodland South American passerines varied from 0.09 to 0.50 ha (Fernández-Juricic et al. 2004b).

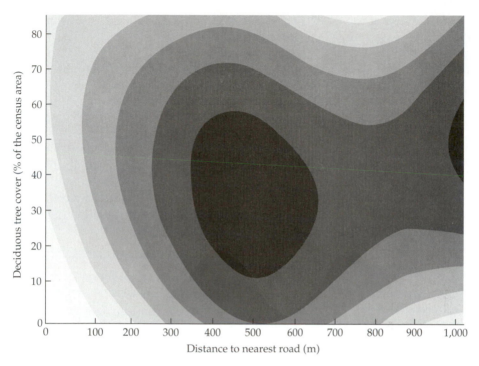

FIGURE 12.3   Abudance of birds in Madrid represented as changes in isodensity bands (the darker the band, the higher the density) in relation to the cover of deciduous trees and the distance to the nearest road. (After Palomino & Carrascal 2007.)

Methods Used to Estimate Patch Buffer Areas

Several methods have been proposed to estimate patch buffer areas (reviewed in Fernández-Juricic et al. 2005c). We show numerical examples below with five of the methods that estimate minimum approaching distances (MADs) and patch buffer areas. These methods require the measurement in the field of alert distance (AD, distance between observer and animal when the latter becomes alert, a proxy of detection) and flight initiation distance (FID, distance between observer and animal when the latter flees) for the target species.

Fernández-Juricic et al. (2009a) studied the tolerance of Belding's Savannah sparrows (*Passerculus sandwichensis beldingi*), an endangered species in the state of California, to recreational activities. This species inhabits only salt marshes, which are declining because of urban sprawl. Taking into consideration recreational approaches made during the breeding season in the Anaheim Bay salt marsh, Fernández-Juricic et al. (2009a) measured alert distance (mean ± standard deviation [SD], AD = 26.48 ± 9.86) and flight initiation distance (mean ± SD, FID = 18.38 ± 7.21). Based on these data, we illustrate how to calculate minimum approaching distances and patch buffer areas with five different methods reviewed in Fernández-Juricic et al. (2005).

METHOD 1 (M1) This method is based on Stalmaster & Newman (1978), McGarigal et al. (1991), Anthony et al. (1995), and Swarthout & Steidl (2001). We plot the cumulative percentage of fleeing individuals against AD and FID to determine the point at which 95% of the individuals become alert ($M1_{AD}$) and take flight ($M1_{FID}$), which can be considered estimates of MAD (Figure 12.4). Based on **Figure 12.4**, the minimum approaching distance causing an alert, MAD_{AD} = 46 m; and the minimum approaching distance causing flight, MAD_{FID} = 31 m.

By taking minimum approaching distance as the radius of a circle, we can calculate a buffer area based on the alert distance or the flight initiation distance: $\pi \times MAD_{AD}^2$ or $\pi \times MAD_{FID}^2$, respectively. Thus,

$$\text{Buffer area } M1_{AD} = \pi \times (46 \text{ m})^2 = 6{,}647.61 \text{ m}^2 = 0.665 \text{ ha}$$

$$\text{Buffer area } M1_{FID} = \pi \times (31 \text{ m})^2 = 3{,}019.07 \text{ m}^2 = 0.302 \text{ ha}$$

METHOD 2 (M2) This method is based on Fox & Madsen (1997): $MAD = 1.5 \times \overline{FID}$, where $\overline{FID}$ is the mean FID.

$$MAD = 1.5 \times 18.38 \text{ m} = 27.57 \text{ m}$$

Buffer areas are calculated as $\pi \times (MAD)^2$. Thus,

$$\text{Buffer area } M2 = \pi \times (27.57 \text{ m})^2 = 2{,}387 \text{ m}^2 = 0.239 \text{ ha}$$

METHOD 3 (M3) This method is based on Rodgers & Smith (1995, 1997) and Rodgers & Schwikert (2002). $MAD = (FID + 1.6495SD) + AD$, where $\overline{FID}$ is the mean FID, SD is the standard deviation of $\overline{FID}$, and $\overline{AD}$ is the mean AD. Thus,

$$MAD = (18.38 + 1.6495 \times 7.21) + 26.48 = 56.76 \text{ m}$$

Buffer areas are calculated as $\pi \times MAD^2$.

$$\text{Buffer area } M3 = \pi \times (56.76 \text{ m})^2 = 10{,}119.87 \text{ m}^2 = 1.01 \text{ ha}$$

METHOD 4 (M4) This method is based on Fernández-Juricic et al. (2001a). $MAD = \overline{AD}$, where $\overline{AD}$ is the mean AD. Thus,

$$MAD = 26.48 \text{ m}$$

MAD is taken as the radius of a circle, so buffer areas can be estimated as $\pi \times MAD^2$.

$$\text{Buffer area } M4 = \pi \times (26.48 \text{ m})^2 = 2{,}203.31 \text{ m}^2 = 0.22 \text{ ha}$$

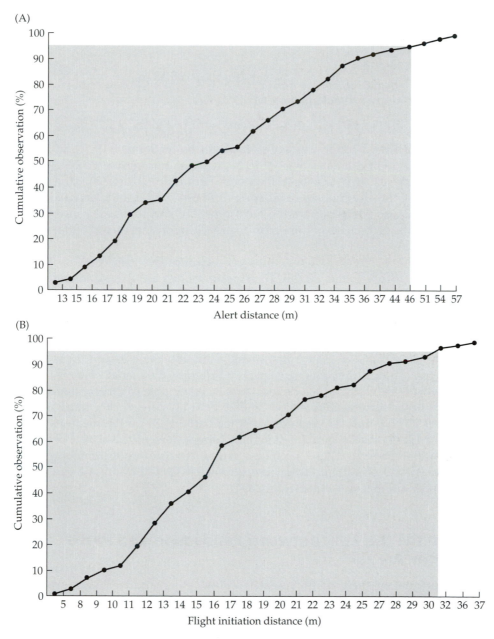

FIGURE 12.4   Cumulative percentage of observations of Belding's Savannah sparrows in relation to (A) alert distance and (B) flight initiation distance. The alert and flight initiation distances corresponding to the 95% level of observations are taken as estimates of minimum approaching distances for Method 1 for estimating buffer areas.

METHOD 5 (M5) This method is based on Vos et al. (1985). Minimum approaching distances correspond to the maximum FID recorded + $\overline{AD}$, where $\overline{AD}$ is the mean AD. Thus,

$$MAD = 37 \text{ m} + 26.48 \text{ m} = 63.48 \text{ m}$$

Buffer areas are estimated as $\pi \times MAD^2$.

$$\text{Buffer area M5} = \pi \times (63.48 \text{ m})^2 = 12{,}660.81 \text{ m}^2 = 1.27 \text{ ha}$$

The average buffer area across methods (considering $M1_{AD}$ and $M1_{FID}$ as two different estimates) was 0.618 ha, with a range between 0.22 and 1.27 ha. Which is then the best method? Fernández-Juricic et al. (2005c) addressed this question by calculating different indices and concluded that Method 2 and Method 3 are the most sensitive and the most conservative ones for the estimation of minimum approaching distances and buffer areas, respectively.

Different methods have different assumptions (see the review in Fernández-Juricic et al. 2005c). Three assumptions are shared among them: (1) the probabilities of fleeing/becoming alert from disturbance are equal in all directions around an individual at any given moment, (2) habitat quality is homogeneous throughout the system, and (3) an individual's use of the buffer area is constant and equal to the carrying capacity of the system. All methods estimate circular buffer areas. Although these assumptions facilitate calculations, they lack realism. For instance, the visual systems of different species would preclude individuals' being able to detect objects with equal probability in all directions, methods rarely consider the natural variability in FID and AD, and most methods are based on the mean value of AD or FID instead of a calculation of the upper percentile estimate of one of these parameters. These problems could lead to underestimating the required size of a buffer area. Future research should address these shortcomings and provide better estimates.

What Is the Link between Landscape and Patch Buffer Areas?

Unfortunately, there is little research on how patch and landscape areas are associated. One can assume that landscape buffer areas will be larger than patch buffer areas by definition, as the latter refers to responses of individual animals. A very simplistic bottom-up approach to link them would be to use the concept of the home range. A **home range** defines the area an animal uses during its daily activities. At any given point throughout the home range, an individual has a patch buffer area around it (**Figure 12.5**). We can then take the home range size as the landscape buffer area of an individual, because if an individual is around the center of its home range, then the patch buffer area is enclosed by the home range. However, if an individual is at the very edge of its home range, its landscape buffer area needs to be

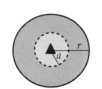  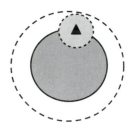

Individual at the center   Individual at the periphery   Home range accounting for extra
   of its home range   of its home range   buffer area at the periphery

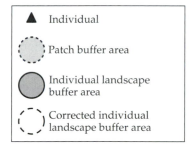

▲ Individual

⋯ Patch buffer area

⬤ Individual landscape
    buffer area

⟨⟩ Corrected individual
    landscape buffer area

FIGURE 12.5   Schematic representation of the patch buffer area and home range of an individual animal (represented by a triangle). Abbreviations: *a*, radius of the buffer area; *r*, radius of the home range. Home ranges are larger than patch buffer areas. When the individual is at the center of its home range, its patch buffer area is included with the home range area. Under this condition, the home range can be considered similar to the landscape buffer area of this individual. However, when the individual is at the periphery of its home range, a correction needs to be calculated to account for the extra buffer area.

expanded to include an area corresponding to the radius of the patch buffer area (see Figure 12.5). We call this parameter the *corrected individual landscape buffer area*.

Here is the simple math. For the sake of simplicity, let's assume that the home range is of circular shape, with radius *r*, and that the radius of the patch buffer area is *a*. The area of the corrected individual landscape buffer area (home rage plus patch buffer area) is:

$$\text{Corrected individual landscape buffer area} = \pi \times (r + a)^2$$

And the difference between the home rage area and the corrected individual landscape buffer area can be estimated as: $\pi \times a(2r+a^2)$.

For instance, imagine that the radius of the home range is $r = 2$ km (home range = 12.56 km^2), and the radius of the patch buffer area (or minimum approaching distance) is $a = 0.030$ km (30 m). Following the equations presented above, the corrected individual landscape buffer area will be 12.94 km^2, with an added area of 0.38 km^2, which represents a 3.03% increase in area.

The bottom line of this simple exercise is that, conceptually, the individual landscape buffer area needs to include both the area of the home range and an additional area (the patch buffer area) to account for situations when the individual is at the periphery of its home range.

What would be the total area necessary to protect the population of this species? If we calculate the corrected individual landscape buffer area, one very simplistic approach is to multiply this area by the number of individuals in the population. This calculation makes many assumptions (e.g., homogeneous habitat, regular food availability, no spatial overlap between individuals) and can be applied only to territorial species. If you are trying to manage a colonial species, then the colony size augmented by the patch buffer areas for the individuals nesting at the periphery of the colony is a reasonable way of calculating the overall protected area. This can be done by following the approach described above to calculate the corrected individual landscape buffer area. However, this estimate does not consider how the individuals forage, which may become important if they must leave the breeding colony to seek food and particularly if there is disturbance of the feeding areas.

We acknowledge that the approach we presented here has conceptual value only. Novel methods to estimate buffer areas are sorely needed, with the goal of developing better management tools for the local protection of species of conservation concern. This is an area that conservation behavior has the potential to make meaningful contributions to in the future.

Buffer Areas Upside Down: Repelling Wildlife

Patch buffer areas can be useful in limiting the disturbance to wildlife through a reduction in avoidance behavior by birds. However, some management problems require that wildlife be repelled locally to reduce interactions between wildlife and humans. For instance, we may want to repel birds from cultivated areas to minimize crop loss, or we may want to reduce bird presence at airports to reduce the chances of bird strikes (e.g., the January 2009 US Airways crash landing in the Hudson River was the result of a strike with Canada geese (*Branta canadensis*; Marra et al. 2009). We can then reverse the concept of buffer areas in order to estimate areas that will be avoided by wildlife as a result of some active repellent. For instance, if we set up a plastic owl model in the middle of a field, there are several questions that will require answering. First, what is the active area around the owl model that will be avoided by birds? Second, how long will the avoidance behavior last? Third, based on the answer to the first question, how many owl models do we need to set up to repel birds from a larger area? Fourth, how do these owl models need to be spaced out?

Wildlife biologists use a wide variety of scaring and hazing devices (cannons, predator models, lasers, etc.). Answering the four aforementioned questions using behavioral approaches can shed light on the effectiveness of these devices for different species under different ecological conditions.

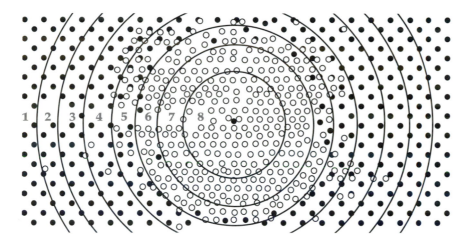

FIGURE 12.6   Distribution of food cups with soybeans in an aviary that was divided into eight areas called blocks. A bird repellent device was positioned at the center of the aviary. Closed circles represented cups where food was consumed, whereas open circles represent food unconsumed by rufous turtle doves. (After Nakamura 1997.)

Nakamura (1997) developed a model and conducted an experiment to estimate the effective areas of devices to scare birds. He built a large aviary (24 × 12.5 m) in which he deposited 1,000 cups containing soybeans and divided the area into eight blocks (**Figure 12.6**). He let rufous turtle doves (*Streptopelia orientalis*) feed in the aviary under conditions with and without a scaring device at the center of the aviary. He used three main types of devices: a mannequin representing a human, a stuffed crow representing a nest predator, and a Gayla kite with two eye spots.

Based on rate of food consumption, Nakamura estimated the rate of stay in a unit block (b_i, i varying from the first and outermost to the eighth and innermost block):

$$b_i = \frac{-D_{max,i}\ln\left(1-f_i\right)}{aN_it - \displaystyle\sum_{k=1}^{i-1}D_{max,k}f_k}$$

$D_{max,i}$ is the amount of food that occupies the whole area of a given block, a is the per capita foraging rate, N_i is the number of birds present in a given block, t is the time interval, $D_{max,k}$ is the amount of food in all the other blocks with the exception of the ith block, f_i is the ratio between the amount of food eaten in the ith block and the total food available in that block, and f_k represents the ratio between the amount of food eaten in all the other blocks and the total amount of food available in all the other blocks but the ith. The staying rate, b_i, is expected to vary with the distance to a device, such that b_i will decrease as it gets closer to it.

Nakamura (1997) defined the effective area of a device as the area within which b_i was reduced to less than half of the b_i under control conditions without the scarer. Nakamura found that the mannequin and the kite had the largest effective areas. Although this experiment was done under semi-natural conditions, the approach can easily be implemented in field conditions, using feeding stations like the ones described for giving-up densities (see Chapter 6). Scaling up the spatial and temporal scales of Nakamura's approach can allow us to establish how the behavioral responses of birds to wildlife repellents can affect the area with little wildlife use and the temporal responses to the repellent (e.g., habituation, sensitization).

Further Reading

Fernández-Juricic et al. (2005c) describes multiple methods to calculate buffer areas. Jarvis (2008) is a comprehensive summary of many studies of human impacts on wildlife.

Afterword

We hope that this book has provided tools that you can use to integrate insights from animal behavior into conserving and managing wildlife.

We believe conservation behavior is a rich subdiscipline with incredible potential. We recognize that many conservation and wildlife management problems are driven by human population size and the associated increase in resource needs. Conservation behavior has little to offer to solve those problems. However, as Sutherland (1998b) wrote in an important paper, "Many conservation projects involve trying to manipulate behaviour." With a focused problem for which we must manipulate an animal's behavior, conservation behavior has actually a lot to offer. Modifying Sutherland's ideas a bit, we may say that if we learn how to manipulate the behavior of a target species (which in itself requires a considerable amount of work), we should have a good chance to solve aspects of some (but not all) conservation and management problems. Most of this *Primer* has focused on developing a mechanistic and evolutionary understanding of behaviors that can be manipulated to different degrees.

More than a decade after the inception of conservation behavior, there is some controversy about how successful it has been (see opposing views in Caro 2007 and Buchholz 2007). Yet, there are still many people interested in learning more about how conservation applications can be incorporated into animal behavior research, and how animal behavior can be incorporated into conservation biology and wildlife management. This makes sense for various reasons.

From a very pragmatic point of view, for instance, some funding agencies require principal investigators to emphasize the broader societal impacts of basic science projects (e.g., conservation applications can be quite straightforward). In many circumstances managers are asked, when possible, to take nonlethal approaches to controlling populations, which requires modifying the *behavior* of individuals on a semipermanent basis. We believe this trend for integrating behavior into conservation and management will grow stronger with time. We thus have a wonderful opportunity to ensure that future investments in conservation behavior have the strongest scientific basis possible so that we make the best informed decisions based on solid theory and empirical results.

Not surprisingly, interest in conservation behavior seems slightly higher in younger scientists (undergraduates and graduate students). One of our goals in writing this *Primer* was both to further inspire this future generation and to provide specific tools they (and others) can now use.

Several professional societies have held conservation behavior symposia to point out the links between disciplines. This has been highly welcome, but we believe future symposia should focus on specific topics to solve problems in thematic areas that sorely need attention. The sections below flesh out our list of thematic areas; it is obviously incomplete, given our inherent biases, but it may help target future research efforts.

Captive Breeding

To recover declined populations, wildlife biologists may have to breed animals in captivity. This is often more easily said than done. Understanding natural mate choice patterns is a first step toward increasing breeding success. While we did not talk about it in this *Primer*, some species are referred to as induced ovulators. Such species require mating to stimulate ovulation. Knowing whether a species is an induced ovulator will influence how animals are paired. Similarly, knowing whether there is reproductive suppression or infanticide in a species in nature (and the mechanisms and functions underlying these) will help us develop rearing conditions that should maximize mating success. Once breeding occurs, knowledge of the degree to which a species must learn survival skills, such as predator recognition and food selection, and the mechanisms underlying the acquisition of these skills should help animals survive both in captivity and later on, following reintroduction.

Translocations and Reintroductions

To recover populations from extinction, managers often must translocate or reintroduce animals from other locations. Some of these management actions are not very successful, because mortality rates are unacceptably high. There is a compelling welfare and management need to reduce such mortality. Wildlife biologists armed with a mechanistic understanding of why animals are attracted to and repelled from certain areas, along with a fundamental understanding of habitat selection and vulnerability to predation, can help reduce this mortality and increase the success of these management actions.

Novel Tools Used to Survey Endangered and Threatened Species

While it is conceptually attractive to use animal sounds for the noninvasive survey or census of natural populations (see Chapter 8), more research is needed regarding the amount of information necessary to make the algorithms distinguish between known individuals and to recognize unknown individuals. For instance, how many vocalizations are necessary to reach a high degree of accuracy? Does this minimum number of vocalizations vary between species with different ecological requirements? How can ambient noise affect the performance of the algorithms, and what are the solutions that will overcome ambient noise effects when species are surveyed in different habitat types? Other automatic systems for surveying species are being developed and implemented at the moment (video cameras, radar, etc.); however, we need more studies that will validate the accuracy of these methods to distinguish between species based on their behavioral parameters (e.g., movement patterns).

Urbanization

The degree of urbanization and the rate of increase in urban sprawl have made us think more about how key ecological processes (pollination, risk of predation, habitat selection) may be affected in human-dominated landscapes and suburban areas. Understanding the behavioral responses of species to different types of human activities or human-modified habitats is essential in predicting whether or not a species will habituate. Ultimately, generating this knowledge across species will promote their conservation and allow us to develop management plans that will enable species to coexist with us in urbanized landscapes. Although correlational research has played a relevant role in urban ecology research, more mechanistic approaches that involve behavior (e.g., Shustack & Rodewald 2008) are sorely needed.

Attracting Animals

In lieu of costly translocations and reintroductions to move animals around, sometimes areas can be made more enticing and thus attract animals naturally. This may be a means of naturally supplementing a population so that it is sufficiently large to reduce the likelihood of stochastic population crashes, or it may lure animals to protected locations (e.g., the black-capped vireo example we discussed in Chapter 10). Wildlife biologists armed with a mechanistic understanding of why animals are attracted to certain areas will play an important role in future management.

Repelling Animals

Animals can interfere with industry (Ronconi & St. Clair 2006) or agriculture, they may create human hazards (e.g., bird collisions with aircraft, or bird strikes; Sodhi 2002), or a given species may interfere with the management of another species (e.g., introduced rats on islands with nesting seabirds). Thus, for a variety of reasons, managers may need to repel animals. Developing effective methods to do so should involve a fundamental understanding of how various species assess predation risk, because by capitalizing on natural fear responses, it is often possible to naturally repel animals. Equally important is how target species perceive different stimuli visually, acoustically, and in other ways, because this knowledge will allow us to develop new methods that exploit their sensory world, making repellents more successful (and eventually reducing the costs from wildlife damage). This is a huge open research area in which wildlife biologists can team up with animal behaviorists and sensory biologists to establish interdisciplinary teams to humanely repel animals.

Reducing Mortality

There are many fronts in which animal behavior can be used to reduce mortality of wild populations. Here are just a few examples.

ROAD MORTALITY Roads bisect natural habitat and many species are killed trying to cross roads. Indeed, some are attracted to roadsides because of enhanced foraging opportunities and microhabitat differences. Developing effective means for reducing road kills is an important conservation goal, and it should benefit from a fundamental understanding of what attracts animals to roadsides and of stimuli that can be used to redirect their attention to alternative routes.

WIND TURBINE MORTALITY Birds and bats are being killed by wind turbines, and we do not really have effective strategies to prevent these unfortunate kills. As wind turbines produce more and more of the world's energy, we expect that these mortalities will increase. By learning why animals are hitting turbines, we can help reduce this source of mortality.

ALTERNATIVE ENERGY DEVELOPMENT In addition to wind turbines, new renewable sources of energy will create a new set of stimuli that animals have never experienced. By developing an understanding of how animals interact behaviorally with these novel stimuli, we should be able to generate effective procedures to reduce mortality and allow coexistence with renewable sources of energy.

FISHERIES BYCATCH *Bycatch* is the term used when nontarget fish, turtles, or marine mammals are incidentally captured (Davies et al. 2008). In many cases, these animals (now dead) are thrown back into the ocean. Fisheries bycatch accounts for an estimated 40% of global marine catches (Davies et al. 2008). There is a compelling need to reduce this mortality. Learning about the behavior of these animals around nets and why they are caught can help target focal species and create novel procedures to avoid bycatch.

The Link between Response and Fitness

Animals may respond in a myriad of ways to anthropogenic stressors (Tarlow & Blumstein 2007). However, simply because there is a response to a human-related stimulus does not mean that an individual's fitness is reduced by that stimulus. Critics often point to this as a failure of conservation behavior, but we believe that it provides an important challenge. While difficult, studies that link behavioral or physiological responses to fitness consequences are essential because they provide the demographic link between anthropogenic stressors and population biology. Importantly, determining the specific nature of the relationship between behavioral responses and fitness is essential, and it need not be linear (e.g., Kight & Swaddle 2007; Busch & Hayward 2009). In this case, by asking questions about the shapes of fitness functions, we may also produce cutting-edge evolutionary research.

Developing Predictive Models of Disturbance

We know that some animals are more likely to be disturbed by humans than others. Why? Developing predictive models of disturbance will fill this gap. We (Blumstein et al. 2005; Blumstein 2006a) have taken a comparative approach to this problem. By studying the evolutionary ecology of antipredator behavior across different species and by using these data to parameterize individual-based models (Bennett et al. 2009), we have illustrated one way of identifying correlates of disturbance. More such comparative studies are needed that focus on other behaviors of conservation relevance (e.g., habitat selection, mating preferences, foraging preferences).

Habituation and Learning

Theories of learning have existed for over a century. Remarkably, we still have much to learn about learning in nature. For instance, Yellowstone National Park has over 3 million visitors a year. Knowledge of how wildlife, such as bears, wolves, and elk, respond to hazing, aversive conditioning, and so forth is key to keeping both the wildlife and the humans safe. Doing so will facilitate wildlife conservation. The same issues are relevant to predator–livestock conflicts. If we wish to coexist with carnivores, we have to develop techniques to keep these species away from livestock, and this requires a deep understanding of how animals learn to respond to aversive stimuli and how they may habituate to repeated presentations of such stimuli.

Furthermore, we still do not really know how the processes of habituation and sensitization evolve. Why, for instance, do some species habituate to human disturbance while others sensitize? Can we use this knowledge to develop novel ways to manage human impacts? Can we use this knowledge to predict which species may be more vulnerable to the long arm of anthropogenic impacts? By developing an evolutionary ecology of learning, similar to what we have begun to develop as an evolutionary ecology of fear (Blumstein 2006a), we will be in a better position to understand why species are particularly vulnerable to disturbance.

Other Effects of Anthropogenic Activities on Behavior

While a growing body of literature addresses how human activities can influence communication (both acoustic and visual), human-related stimuli can also influence other aspects of animal behavior. For instance, sounds may directly modify behavior by repelling animals, masking other biologically important sounds (such as the sounds of predators), and capturing an individual's attention and making it more difficult for the animal to focus on other tasks (such as detecting predators). There is a compelling need to identify how the stimuli humans create modify behavior in a multidimensional way, which means studying the effects of stimuli from different sensory modalities on different behavioral parameters.

Climate Change and Invasive Species

Changes in the world's climate affect the rate at which invasive species disperse (Walther et al. 2009). This creates opportunities for new between-species interactions (predator–prey interactions, competitive interactions, etc.). Many of these interactions take the form of direct encounters between a resident and a novel species. Understanding these direct behavioral interactions may complement other studies that consider the indirect effects of invasions (food competition, preemptive use of common resources, etc.). These direct and indirect sources can be integrated to better predict the outcome of species' interacting in human-modified landscapes. Conservation behavior has ample opportunities to make meaningful contributions while simultaneously developing new ecological insights into species interactions.

Final Thoughts

One of the parameters used to measure the success of a discipline is its ability to generate new knowledge in peer-reviewed journals. However, besides developing novel science, there are other actions that make conservation behavior successful but that are not necessarily published. For instance, already published animal behavior knowledge can be used to solve daily problems, and animal behaviorists can give talks at schools about how the behavior of an endangered or threatened species in response to human activities can lead to human–wildlife problems or even extinction. These and other examples make conservation behavior part of our professional lives, much as habitat selection concepts are part of our broad ecological thinking.

There are many other ways we can promote the use of good animal behavior science in making management or conservation decisions. Maybe with the use of social networking tools, new virtual communities of managers, conservation biologists, and animal behaviorists can be created where they can share specific knowledge from each other, learn about the ample expertise of wildlife biologists, and ask questions about basic behavioral knowledge that can be applied to species of interest. Maybe editors of behavioral journals can encourage authors to explore the conservation implications of animal behavior findings. Or the editors of conservation biology journals can encourage authors to describe the animal behavior theoretical framework used in their studies. We are confident that the new generations will come up with novel ways of implementing conservation behavior.

The Animal Behavior Society's Conservation Committee (www. animalbehavior.org/ABSConservation/ConservationBehaviorist) publishes a newsletter, *The Conservation Behaviorist*, and sponsors workshops, symposia, and round table discussions. The committee welcomes more participants and interdisciplinary input.

We hope that our *Primer* will inspire the future cohort of wildlife biologists who will be at the forefront of solving conservation biology and wildlife management problems, and we hope that the application of such behavioral knowledge will help in different ways save what is left of our precious biodiversity. We are eager to learn about your successes! Please feel free to share with us your conservation behavior stories by emailing us (Dan: marmots@ucla.edu; Esteban: efernan@purdue.edu).

Credits

CHAPTER 1 has drawn upon material from D. T. Blumstein, 2009, Tinbergen's four questions, in *The Encyclopedia of Applied Animal Behavior and Welfare*, D. Mills (ed.), CAB International, Wallingford, UK.

BOX 8.1 draws upon material from D. T. Blumstein, & O. Munos, 2005, Individual, age and sex-specific information is contained in yellow-bellied marmot alarm calls, *Animal Behaviour*, 69, 353–361, with permission from Elsevier.

CHAPTER 10 has drawn upon material from D. T. Blumstein, Social behaviour in conservation, T. Székely, J. Komdeur & A. J. Moore, *Social Behaviour: Genes, Ecology and Evolution*, Cambridge University Press, forthcoming in 2010 and is reproduced here with permission.

PARTS OF CHAPTER 11 are reprinted from L. L. Anthony and D. T. Blumstein, 2000, Integrating behaviour into wildlife conservation: the multiple ways that behaviour can reduce N_e, *Biological Conservation*, 95, 303–315, with permission from Elsevier.

PARTS OF CHAPTER 12 are reprinted or have drawn upon material from E. Fernández-Juricic, P. Venier, D. Renison, & D. T. Blumstein, 2005, Sensitivity of wildlife to spatial patterns of recreationist behavior: A critical assessment of minimum approaching distances and buffer areas for grassland birds, *Biological Conservation*, 125, 225–235, with permission from Elsevier.

Literature Cited

Abrahams, M. V. & Dill, L. M. 1989. A determination of the energetic equivalence of the risk of predation. *Ecology* **70**, 999–1007.

Alberts, A. C., Lemm, J. M., Perry, A. M., Morici, L. A. & Phillips, J. A. 2002. Temporary alteration of local social structure in a threatened population of Cuban iguanas (*Cyclura nubila*). *Behav. Ecol. Sociobiol.* **51**, 324–335.

Alcock, J. 2009. *Animal Behavior*. 9th ed. Sunderland, MA: Sinauer Associates.

Allen, J. A. 1988. Frequency-dependent selection by predators. *Philos. Trans. R. Soc. Lond. B Biol. Sci.* **319**, 485–503.

Allen, J. A. & Weale, M. E. 2005. Anti-apostatic selection by wild birds on quasinatural morphs of the land snail *Cepaea hortensis*: A generalised linear mixed models approach. *Oikos* **108**, 335–343.

Allen, J. A., Raison, H. E. & Weale, M. E. 1998. The influence of density on frequency-dependent selection by wild birds feeding on artificial prey. *Proc. R. Soc. Lond. B Biol. Sci.* **265**, 1031–1035.

Andersson, M. 1994. *Sexual Selection*. Princeton: Princeton University Press.

Anonymous. 1910. Three hundred dollars reward. *N.Y. Times* **16 January 1910**, SM7.

Anthony, L. L. & Blumstein, D. T. 2000. Integrating behaviour into wildlife conservation: The multiple ways that behaviour can reduce N_e. *Biol. Conserv.* **95**, 303–315.

Baerwald, E. F., D'Amours, G. H., Klug, B. J. & Barclay, R. M. R. 2008. Barotrauma is a significant cause of bat fatalities at wind turbines. *Curr. Biol.* **18**, R695–696.

Bainbridge, D. R. J. & Jabbour, H. N. 1998. Potential of assisted breeding techniques for the conservation of endangered mammalian species in captivity: A review. *Vet. Rec.* **143**, 159–168.

Baker, S. E., Ellwood, S. A., Watkins, R. & Macdonald, D. W. 2005. Non-lethal control of wildlife: Using chemical repellents as feeding deterrents for the European badger *Meles meles*. *J. Appl. Ecol.* **42**, 921–931.

Ballinger, K. E., Jr. 2001. Method of deterring birds from plant and structural surfaces. US Patent 6,328,986. www.freepatentsonline.com/6328986.html.

Balmford, A., Beresford, J., Green, J., Naidoo, R., Walpole, M. & Manica, A. 2009. A global perspective on trends in nature-based tourism. *PLoS Biol.* **7(6)**, e1000144.

Baltz, A. P. & Clark, A. B. 1999. Does conspecific attraction affect nest choice in budgerigars (*Melopsittacus undulatus*: Psittacidae: Aves)? *Ethology* **105**, 583–594.

Bangs, E. E. & Fritts, S. H. 1996. Reintroducing the gray wolf to central Idaho and Yellowstone National Park. *Wildl. Soc. Bull.* **24**, 402–413.

Baptista, L. F. & Gaunt, S. L. L. 1997. Bioacoustics as a tool in conservation studies. In *Behavioral approaches to conservation in the wild*, R. Buchholz & J. R. Clemmons (eds.), pp. 212–242. New York: Cambridge University Press.

Bateman, A. J. 1948. Intrasexual selection in *Drosophila*. *Heredity* **2**, 349–368.

Battin, J. 2004. When good animals love bad habitats: Ecological traps and the conservation of animal populations. *Conserv. Biol.* **18**, 1482–1491.

Beale, C. M. & Monaghan, P. 2004. Human disturbance: People as predation-free predators? *J. Appl. Ecol.* **41**, 335–343.

Beauchamp, G. 1998. The effect of group size on mean food intake rate in birds. *Biol. Rev.* **73**, 449–472.

Beauchamp, G. & Fernández-Juricic, E. 2005. The group-size paradox: Effects of learning and patch departure rules. *Behav. Ecol.* **16**, 352–357.

Bednekoff, P. A. & Lima, S. L. 1998. Randomness, chaos and confusion in the study of antipredator vigilance. *Trends Ecol. Evol.* **13**, 284–287.

Beecher, M. D. 1989a. Signalling systems for individual recognition: An information theory approach. *Anim. Behav.* **38**, 248–261.

Beecher, M. D. 1989b. The evolution of parent–offspring recognition in swallows. In *Contemporary issues in comparative psychology*, D. A. Dewsbury (ed.), pp. 360–380. Sunderland, MA: Sinauer Associates.

Beier, P., Majka, D. R. & Spencer, W. D. 2008. Forks in the road: Choices in procedures for designing wildland linkages. *Conserv. Biol.* **22**, 836–851.

Beissinger, S. R. & McCullough, D. R. (eds.) 2002. *Population viability analysis*. Chicago: University of Chicago Press.

Bekoff, M. 1999. Lynx and academic freedom. *Daily Camera* **22 July 1999**, 7A.

Bekoff, M. 2002. The importance of ethics in conservation biology: Let's be ethicists not ostriches. *Endang. Sp. Update* **19**, 23–26.

Bélisle, M. 2005. Measuring landscape connectivity: The challenge of behavioral landscape ecology. *Ecology* **86**, 1988–1995.

Bélisle, M. & St. Clair, C. C. 2001. Cumulative effects of barriers on the movements of forest birds. *Conserv. Ecol.* **5(2)**: 9, www.consecol.org/vol5/iss2/art9.

Bell, A. M. 2004. An endocrine disrupter increases growth and risky behavior in threespined sticklebacks (*Gasterosteus aculeatus*). *Horm. Behav.* **45**, 108–114.

Bennett, V. J., Beard, M., Zollner, P. A., Fernández-Juricic, E., Westphal, L. & LeBlanc, C. L. 2009. Understanding wildlife responses to human disturbance through simulation modelling: A management tool. *Ecol. Complex.* **6**, 113–134.

Berdoy, M., Webster, J. P. & Macdonald, D. W. 2000. Fatal attraction in rats infected with *Toxoplasma gondii*. *Proc. R. Soc. Lond. B Biol. Sci.* **267**, 1591–1594.

Bernstein, C., Krebs, J. R. & Kacelnik, A. 1991. Distribution of birds among habitats: Theory and practice. In *Bird population studies: Relevance to conservation and management*, C. M. Perrins, J.-D. Lebreton & G. J. M. Hirons (eds.), pp. 317–345. Oxford: Oxford University Press.

Bhadra, A., Jordán, F., Sumana, A., Deshpande, S. A. & Gadagkar, R. 2009. A comparative social network analysis of wasp colonies and classrooms: Linking network structure to functioning. *Ecol. Complex.* **6**, 48–55.

Bibby, C. J., Burgess, N. D., Hill, D. A. & Mustoe, S. H. 2000. *Bird census techniques*. 2nd ed. London: Academic Press.

Blackwell, B. F., Fernandez-Juricic, E., Seamans, T. W. & Dolan, T. 2009a. Avian visual system configuration and behavioural response to object approach. *Anim. Behav.* **77**, 673–684.

Blackwell, B. F., DeVault, T. L., Fernández-Juricic, E. & Dolbeer, R. A. 2009b. Wildlife collisions with aircraft: A missing component of land-use planning for airports. *Landsc. Urban Plann.* **93**, 1–9.

Blem, C. R. 1990. Avian energy storage. *Curr. Ornithol.* **7**, 59–113.

Blumstein, D. T. 1992. Multivariate analysis of golden marmot maximum running speed: A new method to study MRS in the field. *Ecology* **73**, 1757–1767.

Blumstein, D. T. 1997. Infanticide among golden marmots (*Marmota caudata aurea*). *Ethol. Ecol. Evol.* **9**, 169–173.

Blumstein, D. T. 1998. Female preferences and effective population size. *Anim. Conserv.* **1**, 173–177.

Blumstein, D. T. 2000. Understanding antipredator behavior for conservation. *The Open Country* **1**, 37–44.

Blumstein, D. T. 2002. Moving to suburbia: Ontogenetic and evolutionary consequences of life on predator-free islands. *J. Biogeogr.* **29**, 685–692.

Blumstein, D. T. 2006a. Developing an evolutionary ecology of fear: How life history and natural history traits affect disturbance tolerance in birds. *Anim. Behav.* **71**, 389–399.

Blumstein, D. T. 2006b. The "multi-predator" hypothesis and the evolutionary persistence of antipredator behaviour. *Ethology* **112**, 209–217.

Blumstein, D. T. 2007a. Darwinian decision making: Putting the adaptive into adaptive management. *Conserv. Biol.* **21**, 552–553.

Blumstein, D. T. 2007b. The evolution, function, and meaning of marmot alarm communication. *Adv. Study Behav.* **37**, 371–400.

Blumstein, D. T. 2009. Tinbergen's four questions. In *The encyclopedia of applied animal behaviour and welfare*, D. Mills (ed.), pp. 605–606. Wallingford, UK: CAB International.

Blumstein, D. T. In Press. Social behaviour in conservation. In *Social behaviour: Genes, ecology and evolution*, T. Szekely, A. J. Moore & J. Komdeur (eds.). Cambridge: Cambridge University Press.

Blumstein, D. T. & Armitage, K. B. 1998. Life history consequences of social complexity: A comparative study of ground-dwelling sciurids. *Behav. Ecol.* **9**, 8–19.

Blumstein, D. T. & Daniel, J. C. 2002. Isolation from mammalian predators differentially affects two congeners. *Behav. Ecol.* **13**, 657–663.

Blumstein, D. T. & Daniel, J. C. 2003a. Developing predictive models of behavior: Do rock-wallabies receive an antipredator benefit from aggregation? *Aust. Mammal.* **25**, 147–154.

Blumstein, D. T. & Daniel, J. C. 2003b. Foraging behavior of three Tasmanian macropodid marsupials in response to present and historical predation threat. *Ecography* **26**, 585–594.

Blumstein, D. T. & Daniel, J. C. 2004. Yellow-bellied marmots discriminate between the alarm calls of individuals and are more responsive to the calls from juveniles. *Anim. Behav.* **68**, 1257–1265.

Blumstein, D. T. & Daniel, J. C. 2005. The loss of anti-predator behaviour following isolation on islands. *Proc. R. Soc. Lond. B Biol. Sci.* **272**, 1663–1668.

Blumstein, D. T. & Daniel, J. C. 2007. *Quantifying behavior the JWatcher way*. Sunderland, MA: Sinauer Associates.

Blumstein, D. T. & Munos, O. 2005. Individual, age and sex-specific information is contained in yellow-bellied marmot alarm calls. *Anim. Behav.* **69**, 353–361.

Blumstein, D. T., Daniel, J. C., Griffin, A. S. & Evans, C. S. 2000. Insular tammar wallabies (*Macropus eugenii*) respond to visual but not acoustic cues from predators. *Behav. Ecol.* **11**, 528–535.

Blumstein, D. T., Daniel, J. C. & Evans, C. S. 2001. Yellow-footed rock-wallaby (*Petrogale xanthopus*) group size effects reflect a trade-off. *Ethology* **107**, 655–664.

Blumstein, D. T., Daniel, J. C. & Springett, B. P. 2004a. A test of the multi-predator hypothesis: Rapid loss of antipredator behavior after 130 years of isolation. *Ethology* **110**, 919–934.

Blumstein, D. T., Runyan, A., Seymour, M., Nicodemus, A., Ozgul, A., Ransler, F., Im, S., Stark, T., Zugmeyer, C. & Daniel, J. C. 2004b. Locomotor ability and wariness in yellow-bellied marmots. *Ethology* **110**, 615–634.

Blumstein, D. T., Fernandez-Juricic, E., Zollner, P. A. & Garity, S. C. 2005. Inter-specific variation in avian responses to human disturbance. *J. Appl. Ecol.* **42**, 943–953.

Blumstein, D. T., Holland, B.-D. & Daniel, J. C. 2006. Predator discrimination and "personality" in captive Vancouver Island marmots (*Marmota vancouverensis*). *Anim. Conserv.* **9**, 274–282.

Blumstein, D. T., Cooley, L., Winternitz, J. & Daniel, J. C. 2008a. Do yellow-bellied marmots respond to predator vocalizations? *Behav. Ecol. Sociobiol.* **62**, 457–468.

Blumstein, D. T., Richardson, D. T., Cooley, L., Winternitz, J. & Daniel, J. C. 2008b. The structure, meaning, and function of yellow-bellied marmot pup screams. *Anim. Behav.* **76**, 1055–1064.

Blumstein, D. T., Wey, T. W. & Tang, K. 2009. A test of the social cohesion hypothesis: Interactive female marmots remain at home. *Proc. R. Soc. Lond. B Biol. Sci.* **207**, 3007–3012.

Blumstein, D. T., Lea, A. J., Olson, L. E., & Martin, J. 2010. Heritability of anti-predatory traits: vigilance and locomotor performance in marmots. *J. Evol. Biol.* **23**, 879–887.

Boag, P. T. & Grant, P. R. 1981. Intense natural selection in a population of Darwin's finches (Geospizinae) in the Galápagos. *Science* **214**, 82–85.

Boake, C. R. B. 1989. Repeatability: Its role in evolutionary studies of mating behavior. *Evol. Ecol.* **3**, 173–182.

Bock, C. E. & Jones, Z. F. 2004. Avian habitat evaluation: Should counting birds count? *Front. Ecol. Environ.* **2**, 403–410.

Bolnick, D. I, Svanback, R., Fordyce, J. A., Yang, L. H., Davis, J. M., Hulsey, C. D. & Forister, M. L. 2003. The ecology of individuals: Incidence and implications of individual specialization. *Am. Nat.* **161**, 1–28.

Boncoraglio, G. & Saino, N. 2007. Habitat structure and the evolution of bird song: A meta-analysis of the evidence for the acoustic adaptation hypothesis. *Funct. Ecol.* **21**, 134–142.

Borenstein, M., Hedges, L. V., Higgins, J. P. T. & Rothstein, H. R. 2009. *Introduction to meta-analysis*. Chichester, West Sussex, UK: John Wiley & Sons.

Borgatti, S. P. 2003. The key player problem. In *Dynamic social network modeling and analysis: Workshop summary and papers*, R. Breiger, K. Carley & P. Pattison (eds.), pp. 241–252. Washington, DC: Committee on Human Factors, National Research Council.

Both, C., Dingemanse, N. J., Drent, P. J. & Tinbergen, J. M. 2005. Pairs of extreme avian personalities have highest reproductive success. *J. Anim. Ecol.* **74**, 667–674.

Bowers, M. A. & Breland, B. 1996. Foraging of gray squirrels on an urban–rural gradient: Use of the GUD to assess anthropogenic impact. *Ecol. Appl.* **6**, 1135–1142.

Bowkett, A. E. 2009. Recent captive-breeding proposals and the return of the ark concept to global species conservation. *Conserv. Biol.* **23**, 773–776.

Bowler, D. E. & Benton, T. G. 2005. Causes and consequences of animal dispersal strategies: Relating individual behaviour to spatial dynamics. *Biol. Rev.* **80**, 205–225.

Bradshaw, G. A., Schore, A. N., Brown, J. L., Poole, J. H. & Moss, C. J. 2005. Elephant breakdown. *Nature* **433**, 807.

Bremner-Harrison, S., Prodohl, P. A. & Elwood, R. W. 2004. Behavioural trait assessment as a release criterion: Boldness predicts early death in a reintroduction programme of captive-bred swift foxes (*Vulpes velox*). *Anim. Conserv.* **7**, 1–8.

Brown, C. & Laland, K. N. 2003. Social learning in fishes: A review. *Fish Fish.* **4**, 280–288.

Brown, J. S. 1988. Patch use as an indicator of habitat preference, predation risk, and competition. *Behav. Ecol. Sociobiol.* **22**, 37–47.

Brown, J. S. 1992. Patch use under predation risk. 1. Models and predictions. *Ann. Zool. Fenn.* **29**, 301–309.

Brown, J. S. & Kotler, B. P. 2007. Foraging and the ecology of fear. In *Foraging,* D. W. Stephens, J. S. Brown & R. C. Ydenberg (eds.), pp. 437–480. Chicago: University of Chicago Press.

Brown, J. S., Kotler, B. P. & Mitchell, W. A. 1994. Foraging theory, patch use, and the structure of a Negev desert granivore community. *Ecology* **75**, 2286–2300.

Brown, L. 1981. Patterns of female choice in mottled sculpins (Cottidae, Teleostei). *Anim. Behav.* **29**, 375–382.

Brumm, H. 2004. Causes and consequences of song amplitude ajustment in a territorial bird: A case study in nightingales. *An. Acad. Bras. Cienc.* **76**, 289–295.

Brumm, H., Voss, K., Köllmer, I. & Todt, D. 2004. Acoustic communication in noise: Regulation of call characteristics in a New World monkey. *J. Exp. Biol.* **207**, 443–448.

Bryant, A. A. & Page, R. E. 2005. Timing and causes of mortality in the endangered Vancouver Island marmot (*Marmota vancouverensis*). *Can. J. Zool.* **83**, 674–682.

Buchholz, R. 2007. Behavioural biology: An effective and relevant conservation tool. *Trends Ecol. Evol.* **22**, 401–407.

Buchholz, R. & Clemmons, J. R. 1997. Behavioral variation: A valuable but neglected biodiversity. In *Behavioral approaches to conservation in the wild,* J. R. Clemmons & R. Buchholz (eds.), pp. 181–208. Cambridge: Cambridge University Press.

Bull, J. J. 1980. Sex determination in reptiles. *Q. Rev. Biol.* **55**, 3–21.

Burger, J. 1994. The effect of human disturbance on foraging behavior and habitat use in piping plover (*Chardrius melodus*). *Estuaries* **17**, 695–701.

Busch, D. S. & Hayward, L. S. 2009. Stress in a conservation context: A discussion of glucocorticoid actions and how levels change with conservation-relevant variables. *Biol. Conserv.* **142**, 2844–2853.

Byers, J. A. 1997. *American pronghorn: Social adaptations and the ghosts of predators past.* Chicago: University of Chicago Press.

Cadiou, B., Monnat, J.-Y. & Danchin, E. 1994. Prospecting in the kittiwake, *Rissa tridactyla*: Different behavioural patterns and the role of squatting in recruitment. *Anim. Behav.* **47**, 847–856.

Campbell, D. L. & Bullard, R. W. 1972. A preference-testing system for evaluating repellents for black-tailed deer. *Proc. Vertebr. Pest Conf.* **5**, 56–63.

Campbell, D. T. & Stanley, J. C. 1963. *Experimental and quasi-experimental designs for research.* Chicago: Rand McNally College Publishing Company.

Campomizzi, A. J., Butcher, J. A., Farrell, S. L., Snelgrove, A. G., Collier, B. A., Gutzwiller, K. J., Morrison, M. L. & Wilkins, R. N. 2008. Conspecific attraction is a missing component in wildlife habitat modeling. *J. Wildl. Manag.* **72**, 331–336.

Carbone, C. & Gittleman, J. L. 2002. A common rule for the scaling of carnivore density. *Science* **295**, 2273–2276.

Carbone, C., Cowlishaw, G., Isaac, N. J. B. & Rowcliffe, J. M. 2005. How far do animals go? Determinants of day range in mammals. *Am. Nat.* **165**, 290–297.

Carbone, C., Teacher, A. & Rowcliffe, J. M. 2007. The costs of carnivory. *PLoS Biol.* **5(2)**, e22.

Carlstead, K., Mellen, J. & Kleiman, D. G. 1999. Black rhinoceros (*Diceros bicornis*) in U.S. zoos: I. Individual behavior profiles and their relationship to breeding success. *Zoo Biol.* **18**, 17–34.

Caro, T. 2005. *Antipredator defenses in birds and mammals.* Chicago: University of Chicago Press.

Caro, T. 2007. Behavior and conservation: A bridge too far? *Trends Ecol. Evol.* **22**, 394–400.

Caro, T. M. 1986a. The functions of stotting: A review of the hypotheses. *Anim. Behav.* **34**, 649–662.

Caro, T. M. 1986b. The functions of stotting in Thomson's gazelles: Some tests of the predictions. *Anim. Behav.* **34**, 663–684.

Caro, T. M. (ed.) 1998. *Behavioral ecology and conservation biology.* Cambridge: Cambridge University Press.

Caro, T. M. & O'Doherty, G. 1999. On the use of surrogate species in conservation biology. *Conserv. Biol.* **13**, 805–814.

Caro, T. M., Eadie, J. & Sih, A. 2005. Use of substitute species in conservation biology. *Conserv. Biol.* **19**, 1821–1826.

Caro, T. M., Young, C. R., Cauldwell, A. E. & Brown, D. E. E. 2009. Animal breeding systems and big game hunting: Models and application. *Biol. Conserv.* **142**, 909–929.

Cheney, D. L. & Seyfarth, R. M. 1980. Vocal recognition in free-ranging vervet monkeys. *Anim. Behav.* **28**, 362–367.

Cheney, D. L. & Seyfarth, R. M. 1990. *How monkeys see the world.* Chicago: University of Chicago Press.

Chesser, R. K. 1991a. Gene diversity and female philopatry. *Genetics* **127**, 437–447.

Chesser, R. K. 1991b. Influence of gene flow and breeding tactics on gene diversity within populations. *Genetics* **129**, 573–583.

Chesser, R. K., Sugg, D. W., Rhodes, O. E., Jr. & Smith, M. H. 1996. Gene conservation. In *Population processes in ecological space and time,* O. E. Rhodes Jr., R. H. Cheser & M. H. Smith (eds.), pp. 237–252. Chicago: University of Chicago Press.

Chetkiewicz, C.-L. B., St. Clair, C. C., & Boyce, M. S. 2006. Corridors for conservation: Integrating pattern and process. *Annu. Rev. Ecol. Evol. Syst.* **37**, 317–342.

Chiszar, D., Krauss, S., Shipley, B., Trout, T. & Smith, H. M. 2009. Response of hatchling Komodo dragons (*Varanus komodoensis*) at Denver Zoo to visual and chemical cues arising from prey. *Zoo Biol.* **28**, 29–34.

Christ, C., Hillel, O., Matus, S. & Sweeting, J. 2003. *Tourism and biodiversity: Mapping tourism's global footprint.* Washington, DC: Conservation International.

Clark, C. W. & Mangel, M. 1986. The evolutionary advantages of group foraging. *Theor. Popul. Biol.* **30**, 45–79.

Clemmons, J. R. & Buchholz, R. (eds.) 1997. *Behavioral approaches to conservation in the wild.* Cambridge: Cambridge University Press.

Clobert, J., de Fraipont, M. & Danchin, É. 2008. Evolution of dispersal. In *Behavioral ecology,* É. Danchin, L.-A. Giraldeau & F. Cézilly (eds.), pp. 323–359. Oxford: Oxford University Press.

Clotfelter, E. D., Bell, A. M. & Levering, K. R. 2004. The role of animal behaviour in the study of endocrine-disrupting chemicals. *Anim. Behav.* **68**, 465–476.

Cohen, J. 1988. *Statistical power analysis for the behavioral sciences.* 2nd ed. Hillsdale, NJ: Lawrence Erlbaum Associates.

Coleman, A., Richardson, D., Schechter, R. & Blumstein, D. T. 2008. Does habituation to humans influence predator discrimination in Gunther's dik-diks (*Madoqua guentheri*)? *Biol. Lett.* **4**, 250–252.

Coltman, D. W. & Slate, J. 2003. Microsatellite measures of inbreeding: A meta-analysis. *Evolution* **57**, 971–983.

Conner, D. A. 1985. The function of the pika short call in individual recognition. *Z. Tierpsychol.* **67**, 131–143.

Conover, M. R. 2002. *Resolving human–wildlife conflicts: The science of wildlife damage management.* Boca Raton, FL: Lewis Publishers.

Coss, R. G. 1999. Effects of relaxed natural selection on the evolution of behavior. In *Geographic variation in behavior: Perspectives on evolutionary mechanisms,* S. A. Foster & J. A. Endler (eds.), pp. 180–208. Oxford: Oxford University Press.

Coss, R. G. & Goldthwaite, R. O. 1995. The persistence of old designs for perception. *Perspect. Ethol.* **11**, 83–148.

Courchamp, F., Clutton-Brock, T. & Grenfell, B. 1999. Inverse density dependence and the Allee effect. *Trends Ecol. Evol.* **14**, 405–410.

Courchamp, F., Berec, L. & Gascoigne, J. 2008. *Allee effects in ecology and conservation.* Oxford: Oxford University Press.

Croft, D. P., James, R. & Krause, J. 2008. *Exploring animal social networks.* Princeton: Princeton University Press.

Crooks, K. R. & Soulé, M. E. 1999. Mesopredator release and avifaunal extinctions in a fragmented system. *Nature* **400**, 563–566.

Crooks, K. R. & Van Vuren, D. 1995. Resource utilization by two insular endemic mammalian carnivores, the island fox and the island spotted skunk. *Oecologia* **104**, 301–307.

Crowder, L. B., Squires, D. D. & Rice, J. A. 1997. Nonadditive effects of terrestrial and aquatic predators on juvenile estuarine fish. *Ecology* **78**, 1796–1804.

Curio, E. 1966. How finches react to predators. *Animals* **9**, 142–143.

Cuthill, I. C. 2006. Color perception. In *Bird coloration.* Vol. 1. *Mechanisms and measurement,* G. E. Hill & K. J. McGraw (eds.), pp. 3–40. Cambridge: Harvard University Press.

Danchin, E., Boulinier, T. & Massot, M. 1998. Conspecific reproductive success and breeding habitat selection: Implications for the study of coloniality. *Ecology* **79**, 2415–2428.

Daniel, J. C. & Blumstein, D. T. 1998. A test of the acoustic adaptation hypothesis in four species of marmots. *Anim. Behav.* **56**, 1517–1528.

Darwin, C. 1871. *The descent of man, and selection in relation to sex.* Princeton: Princeton University Press.

Davies, R. W. D., Cripps, S. J., Nickson, A. & Porter, G. 2008. Defining and estimating global marine fisheries bycatch. *Mar. Policy* **33**, 661–672.

Davis, M. B. & Shaw, R. G. 2001. Range shifts and adaptive responses to quaternary climate change. *Science* **292**, 673–679.

de la Torre, S., Snowdon, C. T. & Bejarano, M. 2000. Effects of human activities on wild pygmy marmosets in Ecuadorian Amazonia. *Biol. Conserv.* **94**, 153–163.

Dell'Omo, G. (ed.) 2002. *Behavioral ecotoxicology.* New York: John Wiley.

Desrochers, A. & Hannon, S. J. 1997. Gap crossing decisions by forest songbirds during the post-fledging period. *Conserv. Biol.* **11**, 1204–1210.

Desrochers, A., Hannon, S. J., Bélisle, M. & St. Clair, C. C. 1999. Movement of songbirds in fragmented forests: Can we "scale up" from behaviour to explain occupancy patterns in the landscape? In *Proceedings of the 22nd International Ornithological Congress*, pp. 2447–2464. Durban, South Africa. NHBS, www.nhbs.com.

Díaz-Uriarte, R. 2002. Incorrect analysis of crossover trials in animal behaviour research. *Anim. Behav.* **63**, 815–822.

Dobson, F. S. 2007. Gene dynamics and social behavior. In *Rodent societies: An ecological and evolutionary perspective*, J. O. Wolff & P. W. Sherman (eds.), pp. 163–172. Chicago: University of Chicago Press.

Dobson, F. S. & Zinner, B. 2003. Social groups, genetic structure, and conservation. In *Animal behavior and wildlife conservation*, M. Festa-Bianchet & M. Apollonio (eds.), pp. 211–228. Washington, DC: Island Press.

Dobson, F. S., Chesser, R. K., Hoogland, J. L., Sugg, D. W. & Foltz, D. W. 1997. Do black-tailed prairie dogs minimize inbreeding? *Evolution* **51**, 970–978.

Dobson, F. S., Chesser, R. K., Hoogland, J. L., Sugg, D. W. & Foltz, D. W. 1998. Breeding groups and gene dynamics in a socially structured population of prairie dogs. *J. Mammal.* **79**, 671–680.

Dobson, F. S., Chesser, R. K., Hoogland, J. L., Sugg, D. W. & Foltz, D. W. 2004. The influence of social breeding groups on effective population size in black-tailed prairie dogs. *J. Mammal.* **85**, 58–66.

Dodge, R. I. 1877. *The plains of the great west and their inhabitants* New York: G. P. Putnam's Sons.

Dohm, M. R. 2002. Repeatability estimates do not always set an upper limit to heritability. *Funct. Ecol.* **16**, 273–280.

Donahue, M. J. 2006. Allee effects and conspecific cueing jointly lead to conspecific attraction. *Oecologia* **149**, 33–43.

Dugatkin, L. A. 1992. Sexual selection and imitation: Females copy the mate choice of others. *Am. Nat.* **139**, 1384–1389.

Ebensperger, L. A. & Blumstein, D. T. 2006. Sociality in New World hystricognath rodents is linked to predators and burrow digging. *Behav. Ecol.* **17**, 410–418.

Ebensperger, L. A. & Blumstein, D. T. 2007. Functions of non-parental infanticide in rodents. In *Rodent societies*, J. O. Wolff & P. W. Sherman (eds.), pp. 267–279. Chicago: University of Chicago Press.

Egnor, S. E. R. & Hauser, M. D. 2006. Noise-induced modulation in cotton-top tamarins (*Saguinus oedipus*). *Am. J. Primatol.* **68**, 1183–1190.

Elgar, M. A. 1989. Predator vigilance and group size in mammals and birds: A critical review of the empirical evidence. *Biol. Rev.* **64**, 13–33.

Emlen, S. T. & Oring, L. W. 1977. Ecology, sexual selection, and the evolution of mating systems. *Science* **197**, 215–222.

Endler, J. A. 1995. Multiple-trait coevolution and environmental gradients in guppies. *Trends Ecol. Evol.* **10**, 22–29.

Evans, C. S. 1997. Referential communication. *Perspect. Ethol.* **12**, 99–143.

Evans, C. S., Evans, L. & Marler, P. 1993. On the meaning of alarm calls: Functional reference in an avian vocal system. *Anim. Behav.* **46**, 23–38.

Falconer, D. S. 1989. *Introduction to quantitative genetics.* 3rd ed. Hong Kong: Longman Scientific and Technical.

Falconer, D. S. & Mackay, T. F. C. 1996. *Introduction to quantitative genetics.* 4th ed. Harlow, Essex, UK: Longmans Green.

Faulkes, C. G., Bennett, N. C., Bruford, M. W., O'Brien, H. P., Aguilar, G. H. & Jarvis, J. U. M. 1997. Ecological constraints drive social evolution in the African mole-rats. *Proc. R. Soc. Lond. B Biol. Sci.* **264**, 1619–1627.

Fay, R. R. 1988. *Hearing in vertebrates: A psychophysics databook.* Winnetka, IL: Hill-Fay Associates.

Felsenstein, J. 2004. *Inferring phylogenies.* Sunderland, MA: Sinauer Associates.

Fernandez-Duque, E. & Valeggia, C. 1994. Meta-analysis: A valuable tool in conservation research. *Conserv. Biol.* **8**, 555–561.

Fernández-Juricic, E. 2000. Bird community composition patterns in urban parks of Madrid: The role of age, size and isolation. *Ecol. Res.* **15**, 373–383.

Fernández-Juricic, E. 2001. Avian spatial segregation at edges and interiors of urban parks in Madrid, Spain. *Biodivers. Conserv.* **10**, 1303–1316.

Fernández-Juricic, E. 2002. Can human disturbance promote nestedness? A case study with birds in an urban fragmented landscape. *Oecologia* **131**, 269–278.

Fernández-Juricic, E. & Tellería, J. L. 2000. Effects of human disturbance on spatial and temporal feeding patterns of blackbird *Turdus merula* in urban parks in Madrid, Spain. *Bird Study* **47**, 13–21.

Fernández-Juricic, E. & Schroeder, N. 2003. Do variations in scanning behaviour affect tolerance to human disturbance? *Appl. Anim. Behav. Sci.* **84**, 219–234.

Fernández-Juricic, E., Jimenez, M. D. & Lucas, E. 2001a. Alert distance as an alternative measure of bird tolerance to human disturbance: Implications for park design. *Environ. Conserv.* **28**, 263–269.

Fernández-Juricic, E., Sanz, R. & Sallent, E. A. 2001b. Frequency-dependent predation by birds at edges and interiors of woodland. *Biol. J. Linn. Soc.* **73**, 43–49.

Fernández-Juricic, E., Sallent, A., Sanz, R. & Rodriguez-Prieto, I. 2003. Testing the risk-disturbance hypothesis in a fragmented landscape: Nonlinear responses of house sparrows to humans. *Condor* **105**, 316–326.

Fernández-Juricic, E., Vaca, R. & Schroeder, N. 2004a. Spatial and temporal responses of forest birds to human approaches in a protected area and implications for two management strategies. *Biol. Conserv.* **117**, 407–416.

Fernández-Juricic, E., Erichsen, J. T. & Kacelnik, A. 2004b. Visual perception and social foraging in birds. *Trends Ecol. Evol.* **19**, 25–31.

Fernández-Juricic, E., Smith, R. & Kacelnik, A. 2005a. Increasing the costs of conspecific scanning in socially foraging starlings affects vigilance and foraging behaviour. *Anim. Behav.* **69**, 73–81.

Fernández-Juricic, E., Poston, R., De Collibus, K., Morgan, T., Bastain, B., Martin, C., Jones, K. & Treminio, R. 2005b. Microhabitat selection and singing behavior patterns of male house finches (*Carpodacus mexicanus*) in urban parks in a heavily urbanized landscape in the western U.S. *Urban Habitats* **3**, 49–69.

Fernández-Juricic, E., Venier, P., Renison, D. & Blumstein, D. T. 2005c. Sensitivity of wildlife to spatial patterns of recreationist behavior: A critical assessment of minimum approaching distances and buffer areas for grassland birds. *Biol. Conserv.* **125**, 225–235.

Fernández-Juricic, E., Gilak, N., McDonald, J. C., Pithia, P. & Valcarcel, A. 2006. A dynamic method to study the transmission of social foraging information in flocks using robots. *Anim. Behav.* **71**, 901–911.

Fernández-Juricic, E., Beauchamp, G. & Bastain, B. 2007a. Group-size and distance-to-neighbour effects on feeding and vigilance in brown-headed cowbirds. *Anim. Behav.* **73**, 771–778.

Fernández-Juricic, E., Zollner, P. A., LeBlanc, C. & Westphal, L. M. 2007b. Responses of nestling black-crowned night herons (*Nycticorax nycticorax*) to aquatic and terrestrial recreational activities: A manipulative study. *Waterbirds* **30**, 554–565.

Fernández-Juricic, E., Gall, M. D., Dolan, T., Tisdale, V. & Martin, G. R. 2008. The visual fields of two ground foraging birds, house finches and house sparrows, allow for simultaneous foraging and anti-predator vigilance. *Ibis* **150**, 779–787.

Fernández-Juricic, E., Zahn, E., Parker, T. & Stankowich, T. 2009a. California endangered Belding's Savannah sparrow (*Passerculus sandwichensis beldingi*): Tolerance of pedestrian disturbance. *Avian Conserv. Ecol.* **4(2)**:1, www.ace-eco.org/vol4/iss2/art1.

Fernández-Juricic, E., del Nevo, A. & Poston, R. 2009b. Identification of individual and population level variation in vocalizations of the endangered southwestern willow flycatcher (*Empidonax traillii extimus*). *Auk* **126**, 89–99.

Festa-Bianchet, M. & Apollonio, M. (eds.) 2003. *Animal behavior and wildlife conservation*. Washington, DC: Island Press.

Fisher, D. O. & Owens, I. P. F. 2004. The comparative method in conservation biology. *Trends Ecol. Evol.* **19**, 391–398.

Fisher, H. S., Swaisgood, R. R. & Fitch-Snyder, H. 2003. Odor familiarity and female preferences for males in a threatened primate, the pygmy loris *Nycticebus pygmaeus*: Applications for genetic management of small populations. *Naturwissenschaften* **90**, 509–512.

Fisher, H. S., Wong, B. B. M. & Rosenthal, G. G. 2006. Alteration of the chemical environment disrupts communication in a freshwater fish. *Proc. R. Soc. Lond. B Biol. Sci.* **273**, 1187–1193.

Fitch, W. T. & Hauser, M. D. 2002. Unpacking "honesty": Vertebrate vocal production and the evolution of acoustic signals. In *Springer handbook of auditory research,* A. Simmons, R. R. Fay & A. N. Popper (eds.), pp. 65–137. New York: Springer.

Flack, J. C., Girvan, M., de Waal, F. B. M. & Krakauer, D. C. 2006. Policing stabilizes construction of social niches in primates. *Nature* **439**, 426–429.

Foose, T. J. & Ballou, J. D. 1988. Population management: Theory and practice. *Int. Zoo Yearb.* **27**, 26–41.

Foote, A. D., Osborne, R. W. & Hoelzel, A. R. 2004. Whale-call response to masking boat noise. *Nature* **428**, 910.

Forsgren, E. 1992. Predation risk affects mate choice in a gobiid fish. *Am. Nat.* **140**, 1041–1049.

Forsman, J. T., Seppänen, J. & Mönkkönen, M. 2002. Positive fitness consequences of interspecific interaction with a potential competitor. *Proc. R. Soc. Lond. B Biol. Sci.* **269**, 1619–1623.

Foster, S. A. & Endler, J. A. (eds.) 1999. *Geographic variation in behavior: Perspectives on evolutionary mechanisms.* Oxford: Oxford University Press.

Fox, A. D. & Madsen, J. 1997. Behavioural and distributional effects of hunting disturbance on waterbirds in Europe: Implications for refuge design. *J. Appl. Ecol.* **34**, 1–13.

Frankham, R., Ballou, J. D. & Briscoe, D. A. 2002. *Introduction to conservation genetics.* Cambridge: Cambridge University Press.

Fraschetti, S., D'Ambrosio, P., Micheli, F., Pizzolante, F., Bussotti, S. & Terlizzi, A. 2009. Design of marine protected areas in a human-dominated seascape. *Mar. Ecol. Prog. Ser.* **375**, 13–24.

Fretwell, S. D. & Lucas, H. L., Jr. 1970. On territorial behavior and other factors influencing habitat distribution in birds. I. Theoretical development. *Acta Biotheor.* **19**, 16–36.

Frid, A. & Dill, L. M. 2002. Human-caused disturbance stimuli as a form of predation risk. *Conserv. Ecol.* **6**, 11, www.ecologyandsociety. org/issues/view.php?id=11#Synthesis.

Garland, T. G., Jr. & Ives, A. R. 2000. Using the past to predict the present: Confidence intervals for regression equations in phylogenetic comparative methods. *Am. Nat.* **155**, 346–364.

Garland, T. J., Bennett, A. F. & Rezende, E. L. 2005. Phylogenetic approaches in comparative physiology. *J. Exp. Biol.* **208**, 3015–3035.

Ghalambor, C. K. & Martin, T. E. 2000. Parental investment strategies in two species of nuthatch vary with stage-specific predation risk and reproductive effort. *Anim. Behav.* **60**, 263–267.

Gibson, R. M. & Höglund, J. 1992. Copying and sexual selection. *Trends Ecol. Evol.* **7**, 229–232.

Gill, J. A. & Sutherland, W. J. 2000. Predicting the consequences of human disturbance from behavioural decisions. In *Behaviour and conservation,* L. M. Gosling & W. J. Sutherland (eds.), pp. 51–64. Cambridge: Cambridge University Press.

Gill, J. A., Sutherland, W. J. & Watkinson, A. R. 1996. A method to quantify the effects of human disturbance on animal populations. *J. Appl. Ecol.* **33**, 786–792.

Gill, J. A., Norris, K. & Sutherland, W. J. 2001. The effects of disturbance on habitat use by black-tailed godwits *Limosa limosa. J. Appl. Ecol.* **38**, 846–856.

Gilpin, M. E. & Soulé, M. E. 1986. Minimum viable populations: Processes of species extinction. In *Conservation biology: The science of scarcity and diversity,* M. E. Soulé (ed.), pp. 19–34. Sunderland, MA: Sinauer Associates.

Gilroy, J. J. & Sutherland, W. J. 2007. Beyond ecological traps: Perceptual errors and undervalued resources. *Trends Ecol. Evol.* **22**, 351–356.

Giraldeau, L.-A. 1988. The stable group and the determinants of foraging group size. In *The ecology of social behavior,* C. N. Slobodchikoff (ed.), pp. 33–53. San Diego: Academic Press.

Giraldeau, L.-A. & Gillis, D. 1984. Optimal group size can be stable: A reply to Sibly. *Anim. Behav.* **33**, 666–667.

Giraldeau, L.-A., Valone, T. J. & Templeton, J. J. 2002. Potential disadvantages of using socially acquired information. *Philos. Trans. R. Soc. Lond. B Biol. Sci.* **357**, 1559–1566.

Gosling, L. M. & Sutherland, W. J. 2000. *Behaviour and conservation.* Cambridge: Cambridge University Press.

Gosling, S. D. 1998. Personality dimensions in spotted hyenas (*Crocuta crocuta*). *J. Comp. Psychol.* **112**, 107–118.

Gosling, S. D. 2001. From mice to men: What can we learn about personality from animal research? *Psychol. Bull.* **127**, 45–86.

Graur, D. & Martin, W. 2004. Reading the entrails of chickens: Molecular timescales of evolution and the illusion of precision. *Trends Genet.* **20**, 80–86.

Greene, C., Umbanhowar, J., Mangel, M. & Caro, T. 1998. Animal breeding systems, hunter selectivity, and consumptive use in wildlife conservation. In *Behavioral ecology and conservation biology,* T. Caro (ed.), pp. 271–305. New York: Oxford University Press.

Griffin, A. S., Blumstein, D. T. & Evans, C. S. 2000. Training captive-bred or translocated animals to avoid predators. *Conserv. Biol.* **14**, 1317–1326.

Griffin, A. S. & Evans, C. S. 2003. Social learning of antipredator behaviour in a marsupial. *Anim. Behav.* **66**, 485–492.

Griffin, A. S., Evans, C. S. & Blumstein, D. T. 2001. Learning specificity in acquired predator recognition. *Anim. Behav.* **62**, 577–589.

Griffin, A. S., Evans, C. S. & Blumstein, D. T. 2002. Selective learning in a marsupial. *Ethology* **108**, 1103–1114.

Griffith, B., Scott, J. M., Carpenter, J. W. & Reed, C. 1989. Translocation as a species conservation tool: Status and strategy. *Science* **245**, 477–480.

Grimm, V. & Railsback, S. F. 2005. *Individual-based modeling and ecology*. Princeton: Princeton University Press.

Grimm, V., Stillman, R. A., Jax, K. & Goss-Custard, J. 2007. Modeling adaptive behavior in event-driven environments. In *Temporal dimensions of landscape ecology*, J. A. Bissonette & I. Storch (eds.), pp. 59–77. New York: Springer.

Groom, M. J., Meffe, G. K. & Carroll, C. R. 2006. *Principles of conservation biology*. 3rd ed. Sunderland, MA: Sinauer Associates.

Gutzwiller, K. J., Wiedenmann, R. T., Clements, K. L. & Anderson, S. H. 1994. Effects of human intrusion on song occurrence and singing consistency in subalpine birds. *Auk* **111**, 28–37.

Hagar, R. & Jones, C. B. (eds.) 2009. *Reproductive skew in vertebrates: Proximate and ultimate causes*. Cambridge: Cambridge University Press.

Hartley, M. J. & Hunter, M. L., Jr. 1998. A meta-analysis of forest cover, edge effects, and artificial nest predation rates. *Conserv. Biol.* **12**, 465–469.

Harvey, P. H. & Pagel, M. D. 1991. *The comparative method in evolutionary biology*. Oxford: Oxford University Press.

Hausfater, G. & Hrdy, S. B. (eds.) 1984. *Infanticide*. New York: Aldine.

Hayssen, V., van Tienhoven, A. & van Tienhoven, A. 1993. *Asdell's patterns of mammalian reproduction: A compendium of species-specific data*. Ithaca, NY: Cornell University Press.

Hayward, T. J. 1997. Classification by multiple-resolution statistical analysis with application to automated recognition of marine mammal sounds. *J. Acoust. Soc. Am.* **101**, 1516–1526.

Hearne, G. W., Berghaier, R. W. & George, D. D. 1996. Evidence for social enhancement of reproduction in two *Eulemur* species. *Zoo Biol.* **15**, 1–12.

Hedges, S. B. & Kumar, S. 2004. Precision of molecular time estimates. *Trends Genet.* **20**, 242–247.

Heffner, R. S., Koay, G. & Heffner, H. E. 2001. Audiograms of five species of rodents: Implications for the evolution of hearing and the encoding of pitch. *Hear. Res.* **157**, 138–152.

Henry, K. S. & Lucas, J. R. 2008. Coevolution of auditory sensitivity and temporal resolution with acoustic signal space in three songbirds. *Anim. Behav.* **76**, 1659–1671.

Higashi, M. & Yamamura, N. 1993. What determines animal group size? Insider–outsider conflict and its resolution. *Am. Nat.* **142**, 553–563.

Hinde, R. A. 1975. Interactions, relationships and social structure in non-human primates. In *Proceedings from the symposia of the Fifth Congress of the International Primatological Society, Nagoya, Japan, August 1974*, S. Kondo, M. Kawai, A. Ehara & S. Kawamura (eds.), pp. 13–24. Tokyo: Japan Science Press.

Hixon, M. A., Pacala, S. W. & Sandin, S. A. 2002. Population regulation: Historical context and contemporary challenges of open vs. closed systems. *Ecology* **83**, 1490–1508.

Höglund, J. 1996. Can mating systems affect local extinction risks? Two examples of lek-breeding waders. *Oikos* **77**, 184–188.

Hrdy, S. B. 1979. Infanticide among animals: A review, classification, and examination of the implications for the reproductive strategies of females. *Ethol. Sociobiol.* **1**, 13–40.

Hunter, J. E. & Schmidt, F. L. 1990. *Methods of meta-analysis: Correcting error and bias in research findings*. Newbury Park, CA: Sage Publications.

Huntingford, F. A. 1976. The relationship between anti-predator behaviour and aggression among conspecifics in the three-spined stickleback, *Gasterosteus aculeatus*. *Anim. Behav.* **24**, 245–260.

Huntley, B., Collingham, Y. C., Green, R. E., Hilton, G. M., Rahbek, C. & Willis, S. G. 2006. Potential impacts of climatic change upon geographical distributions of birds. *Ibis* **148**, 8–28.

Ikuta, L. A. & Blumstein, D. T. 2003. Do fences protect birds from human disturbance? *Biol. Conserv.* **112**, 447–452.

Ims, R. A. 1995. Movement patterns in relation to spatial structures. In *Mosaic landscapes and ecological processes*, L. Hanson, L. Fahrig, & G. Merriam (eds.), pp. 85–109. London: Chapman & Hall.

IUCN. 2002. *IUCN technical guidelines on the management of ex-situ populations for conservation*. Gland, Switzerland: IUCN.

Jarvis, P. J. 2008. *The reactions of animals to human disturbance: An annotated bibliography*. Birmingham, UK: Urban Environment.

Jasny, M., Reynolds, J., Horowitz, C. & Wetzler, A. 2005. *Sounding the depths*. II. *The rising toll of sonar, shipping and industrial ocean noise on marine life*. Washington, DC: Natural Resources Defense Council.

Jenkins, H. E., Woodroffe, R., Donnelly, C. A., Cox, D. R., Johnston, W. T., Bourne, F. J., Cheeseman, C. L., Clifton-Hadley, R. S., Gettinby, G., Gilks, P., Hewinson, R. G., McInerey, J. P. & Morrison, W. I. 2007. Effects of culling on spatial associations of *Mycobacterium bovis* infections in badgers and cattle. *J. Appl. Ecol.* **44**, 897–908.

Jennions, M. D., Bacwell, P. R. Y. & Passmore, N. I. 1995. Repeatability of mate choice: The effect of size in the African painted reed frog, *Hyperolius marmoratus*. *Anim. Behav.* **49**, 181–186.

Johnson, C. 2006. *Australia's mammal extinctions: A 50,000 year history*. Cambridge: Cambridge University Press.

Johnson, D. H. 1980. The comparison of usage and availability measurements for evaluating resource preference. *Ecology* **61**, 65–71.

Johnson, M. D. 2007. Measuring habitat quality: A review. *Condor* **109**, 489–504.

Jones, A. C. & Gosling, S. D. 2005. Temperament and personality in dogs (*Canis familiaris*): A review and evaluation of past research. *Appl. Anim. Behav. Sci.* **95**, 1–53.

Jones, K. A., Krebs, J. R. & Whittingham, M. J. 2007. Vigilance in the third dimension: Head movement not scan duration differs in response to different predator models. *Anim. Behav.* **74**, 1181–1187.

Kight, C. R. & Swaddle, J. P. 2007. Associations of anthropogenic activity and disturbance with fitness metrics of eastern bluebirds (*Sialia sialis*). *Biol. Conserv.* **138**, 189–197.

Kiltie, R. A. 2000. Scaling of visual acuity with body size in mammals and birds. *Funct. Ecol.* **14**, 226–234.

Kirschel, A. N. G., Blumstein, D. T., Cohen, R. E., Buermann, W., Smith, T. B. & Slabbekoorn, H. 2009a. Birdsong tuned to the environment: Green hylia song varies with elevation, tree cover, and noise. *Behav. Ecol.* **20**, 1089–1095.

Kirschel, A. N. G., Earl, D. A., Yao, Y., Escobar, I., Vilches, E., Vallejo, E. E. & Taylor, C. E. 2009b. Using songs to identify individual Mexican antthrush *Formicarius moniliger*: Comparison of four classification methods. *Bioacoustics* **19**, 1–20.

Kleiman, D. G. 1989. Reintroduction of captive mammals for conservation: Guidelines for reintroducing endangered species into the wild. *BioScience* **39**, 152–161.

Kleiman, D. G., Allen, M. E., Thompson, K. V. & Lumpkin, S. (eds.) 1997. *Wild mammals in captivity: Principles and techniques*. Chicago: University of Chicago Press.

Klem, D., Jr. 1990. Collisions between birds and windows: Mortality and prevention. *J. Field Ornithol.* **61**, 120–128.

Knight, R. L. & Skagen, S. K. 1988. Effects of recreational disturbance on birds of prey: A review. In *Proceedings of the Southwest Raptor Managment Symposium and Workshop*, Institute of Wildlife Research (ed.), pp. 335–359. Reston, VA: National Wildlife Federation Science Technical Serial 11.

Knight, R. L. & Temple, S. A. 1995. Wildlife and recreationists: Coexistence through management. In *Wildlife and recreationists: Coexistence through management and research*, R. L. Knight & K. J. Gutzwiller (eds.). Washington, DC: Island Press.

Komdeur, J. 1992. Importance of habitat saturation and territory quality for evolution of cooperative breeding in the Seychelles warbler. *Nature* **358**, 493–495.

Komdeur, J. & Kats, R. K. H. 1999. Predation risk affects trade-off between nest guarding and foraging in Seychelles warblers. *Behav. Ecol.* **10**, 648–658.

Kotler, B. P., Brown, J. S. & Hasson, O. 1991. Factors affecting gerbil foraging behavior and rates of owl predation. *Ecology* **72**, 2249–2260.

Kotler, B. P., Brown, J. S. & Mitchell, W. A. 1993. Environmental factors affecting patch use in two species of gerbilline rodents. *J. Mammal.* **74**, 614–620.

Kramer, D. L. 1985. Are colonies supraoptimal groups? *Anim. Behav.* **33**, 1031–1032.

Kramer, D. L. & Chapman, M. R. 1999. Implications of fish home range size and relocation for marine reserve function. *Env. Biol. Fish.* **55**, 65–79.

Kramer, D. L., Rangeley, R. W. & Chapman, L. J. 1997. Habitat selection: Patterns of spatial distribution from behavioural decisions. In *Behavioural ecology of teleost fishes*, J.-G. J. Godin (ed.), pp. 37–80. Oxford: Oxford University Press.

Krause, J. & Ruxton, G. D. 2002. *Living in groups*. Oxford: Oxford University Press.

Krebs, C. J. 2001. *Ecology: The experimental analysis of distribution and abundance*. 5th ed. San Francisco: Benjamin-Cummings.

Kruuk, L. E. B. 2004. Estimating genetic parameters in natural populations using the "animal model." *Philos. Trans. R. Soc. Lond. B Biol. Sci.* **359**, 873–890.

Kuijper, D. P. J., Cromsigt, J. P. M. G., Churski, M., Adam, B., Jedrzejewska, B. & Jedrzejewski, W. 2009. Do ungulates preferentially feed in forest gaps in European temperate forest? *For. Ecol. Manag.* **258**, 1528–1535.

Lacey, E. A. 2000. Spatial and social systems of subterranean rodents. In *Life underground: The biology of subterranean rodents*, E. A. Lacey, J. L. Patton & G. N. Cameron (eds.), pp. 257–296. Chicago: University of Chicago Press.

Lahti, D., Johnson, N. A., Ajie, B. C., Otto, S. P., Hendry, A. P., Blumstein, D. T., Coss, R. G., Donohue, K. & Foster, S. A. 2009. Relaxed selection in the wild. *Trends Ecol. Evol.* **24**, 487–496.

Laundré, J. W., Hernández, L. & Altendorf, K. B. 2001. Wolves, elk, and bison: Reestablishing the "landscape of fear" in Yellowstone National Park, U.S.A. *Can. J. Zool.* **79**, 1401–1409.

Lawson, R. E., Putman, R. J. & Fielding, A. H. 2000. Individual signatures in scent gland secretions of Eurasian deer. *J. Zool. (Lond.)* **251**, 399–410.

Le Boeuf, B. J. & Reiter, J. 1988. Lifetime reproductive success in northern elephant seals. In *Reproductive success* T. H. Clutton-Brock (ed.), pp. 344–362. Chicago: University of Chicago Press.

Leger, D. W., Berney-Key, S. D. & Sherman, P. W. 1984. Vocalizations of Belding's ground squirrels, *Spermophilus beldingi. Anim. Behav.* **32**, 753–764.

Linnell, J. D. C., Aanes, R., Swenson, J. E., Odden, J. & Smith, M. E. 1997. Translocation of carnivores as a method for managing problem animals: A review. *Biodivers. Conserv.* **6**, 1245–1257.

Lima, S. L. & Bednekoff, P. A. 1999. Temporal variation in danger drives antipredator behavior: The predation risk allocation hypothesis. *Am. Nat.* **153**, 649–659.

Lima, S. L. & Dill, L. M. 1990. Behavioral decisions made under the risk of predation: A review and prospectus. *Can. J. Zool.* **68**, 619–640.

Lombard, E. 1911. Le signe de l'élévation de la voix [The characteristics of the elevation of the voice]. *Annales des Maladies de l'Oreille, du Larynx, du Nez et du Pharynx* **37**, 101–119.

Lord, A., Waas, J. R. & Innes, J. 1997. Effects of human activity on the behaviour of northern New Zealand dotterel *Charadrius obscurus aquilonius* chicks. *Biol. Conserv.* **82**, 15–20.

Lott, D. F. 1991. *Intraspecific variation in the social systems of wild vertebrates*. Cambridge: Cambridge University Press.

Low, T. 1999. *Feral future*. Ringwood, Victoria: Viking.

Lowe, C. G. & Bray, R. N. 2006. Movement and activity patterns. In *The ecology of marine fishes: California and adjacent waters*, L. G. Allen, D. J. Pondella & M. H. Horn (eds.), pp. 524–553. Berkeley: University of California Press.

Lusseau, D. 2003a. Effects of tour boats on the behavior of bottlenose dolphins: Using Markov chains to model anthropogenic impacts. *Conserv. Biol.* **17**, 1785–1793.

Lusseau, D. 2003b. The emergent properties of a dolphin social network. *Biol. Lett.* **270**, S186–S188.

Lusseau, D., Wilson, B., Hammond, P. S., Grellier, K., Durban, J. W., Parsons, K. M., Barton, T. R. & Thompson, P. M. 2005. Quantifying the influence of sociality on population structure in bottlenose dolphins. *J. Anim. Ecol.* **75**, 14–24.

Luttbeg, B. 1996. A comparative Bayes tactic for mate assessment and choice. *Behav. Ecol.* **7**, 451–460.

Machlis, L., Dodd, P. W. & Fentress, J. C. 1985. The pooling fallacy: Problems arising when individuals contribute more than one observation to the data set. *Z. Tierpsychol.* **68**, 201–214.

Maddison, W. P. & Maddison, D. R. 2001. *MacClade: Analysis of phylogeny and character evolution. Version 4.03.* Sunderland, MA: Sinauer Associates.

Maddison, W. P. & Maddison, D. R. 2005. *Mesquite: A modular system for evolutionary analysis. Version 1.06.* www.mesquiteproject.org.

Maher, C. R. & Lott, D. F. 2000. A review of ecological determinants of territoriality within vertebrate species. *Am. Midl. Nat.* **143**, 1–29.

Maldonado-Chaparro, A. & Blumstein, D. T. 2008. Management implications of capybara (*Hydrochoerus hydrochaeris*) social behavior. *Biol. Conserv.* **141**, 1945–1952.

Manly, B. F. G., McDonald, L. L., Thomas, D. L., McDonald, T. L. & Erickson, W. P. 2002. *Resource selection by animals: Statistical design and analysis for field studies.* Dordrecht, Netherlands: Kluwer Academic Press.

Marra, P. P., Dove, C. J., Dolbeer, R., Dahlan, N. F., Heacker, M., Whatton, J. F., Diggs, N. E., France, C. & Henkes, G. A. 2009. Migratory Canada geese cause crash of US Airways flight 1549. *Front. Ecol. Environ.* **7**, 297–301.

Marler, P. 1960. Bird song and mate selection. In *Animal sounds and communication*, W. F. Lanyon & W. Tavolga (eds.), pp. 348–367. Washington, DC: American Institute of Biological Sciences.

Marshall, M. R. & Cooper, R. J. 2004. Territory size of a migrating songbird in response to caterpillar density and foliage structure. *Ecology* **85**, 432–445.

Martin, P. & Bateson, P. 2007. *Measuring behaviour: An introductory guide*. 3rd ed. Cambridge: Cambridge University Press.

Martin, P. S. & Klein, R. G. (eds.) 1984. *Quaternary extinctions: A prehistoric revolution*. Tuscon: University of Arizona Press.

Martins, E. P. 2003. *COMPARE. Version 4.5.* Bloomington, IN: Department of Biology, Indiana University.

Masataka, N. 1993. Effects of experience with live insects on the development of fear of snakes and squirrel monkeys, *Saimiri sciureus*. *Anim. Behav.* **46**, 741–746.

Mason, G. J., Cooper, J. & Clarebrough, C. 2001. Frustrations of fur-farmed mink. *Nature* **410**, 35–36.

McComb, K., Moss, C., Durant, S. M., Baker, L. & Sayialel, S. 2001. Matriarchs as repositories of social knowledge in African elephants. *Science* **292**, 491–494.

McCracken, K. G. & Sheldon, F. H. 1997. Avian vocalizations and phylogenetic signal. *Proc. Natl. Acad. Sci. USA* **94**, 3833–3836.

McGregor, P. K. & Peake, T. M. 1998. The role of individual identification in conservation biology. In *Behavioral ecology and conservation biology*, T. Caro (ed.), pp. 31–55. New York: Oxford University Press.

McGregor, P. K., Peake, T. M. & Gilbert, G. 2000. Communication behavior and conservation. In *Behaviour and conservation*, L. M. Gosling & W. J. Sutherland (eds.), pp. 261–280. Cambridge: Cambridge University Press.

Mech, S. G. & Zollner, P. A. 2002. Using body size to predict perceptual range. *Oikos* **98**, 47–52.

Midford, P. E., Garland, T. J. & Maddison, W. P. 2005. *PDAP Package of Mesquite. Version 1.07.* www.mesquiteproject.org/pdap_mesquite.

Miller, S. G., Knight, R. L. & Miller, C. K. 1998. Influence of recreational trails on breeding bird communities. *Ecol. Appl.* **8**, 162–169.

Ministry of Forests and Range. 2001. *An introductory guide to adaptive management. Appendix 3.* Victoria, BC: Ministry of Forests and Range, Forest Practices Branch. Available from www.gov.bc.ca/for.

Misenhelter, M. D. & Rotenberry, J. T. 2000. Choices and consequences of habitat occupancy and nest site selection in sage sparrows. *Ecology* **81**, 2892–2901.

Møller, A. P. & Thornhill, R. 1998. Bilateral symmetry and sexual selection: A meta-analysis. *Am. Nat.* **151**, 174–192.

Møller, A. P., Nielsen, J. T. & Garamzegi, L. Z. 2008. Risk taking by singing males. *Behav. Ecol.* **19**, 41–53.

Mönkkönen, M. & Forsman, J. T. 2002. Heterospecific attraction among forest birds: A review. *Ornithol. Sci.* **1**, 41–51.

Mönkkönen, M., Helle, P. & Soppela, K. 1990. Numerical and behavioral responses of migrant passerines to experimental manipulations of residents tits (*Parus* spp.): Heterospecific attraction in northern breeding bird communities? *Oecologia* **85**, 218–225.

Mönkkönen, M., Helle, P., Niemi, G. J. & Montgomery, K. 1997. Heterospecific attraction affects community structure and migrant abundances in northern breeding bird communities. *Can. J. Zool.* **75**, 2077–2083.

Morris, D. W. 1987. Tests of density-dependent habitat selection in a patchy environment. *Ecol. Monogr.* **57**, 269–281.

Morris, D. W. 1988. Habitat-dependent population regulation and community structure. *Evol. Ecol.* **2**, 253–269.

Morris, D. W. 1992. Scales and costs of habitat selection in heterogeneous landscapes. *Evol. Ecol.* **6**, 412–432.

Morris, D. W. 1995. Habitat selection in mosaic landscapes. In *Mosaic landscapes and ecological processes*, L. Hansson, L. Fahrig & G. Merriam (eds.), pp. 110–135. London: Chapman and Hall.

Morris, D. W. 2003. Towards an ecological synthesis: A case for habitat selection. *Oecologia* **136**, 1–13.

Morris, D. W. & Davidson, D. L. 2000. Optimally foraging mice match patch use with habitat differences in fitness. *Ecology* **81**, 2061–2066.

Morris, K., Armstrong, R., Orell, P. & Vance, M. 1998. Bouncing back: Western shield update. *Landscope* **14**, 28–35.

Muller, K. L. 1998. The role of conspecifics in habitat settlement in a territorial grasshopper. *Anim. Behav.* **56**, 479–485.

Müllner, A., Linsenmair, K. E. & Wikelski, M. 2004. Exposure to ecotourism reduces survival and affects stress response in hoatzin chicks (*Opisthocomus hoazin*). *Biol. Conserv.* **118**, 549–558.

Nakamura, K. 1997. Estimation of effective area of bird scarers. *J. Wildl. Manag.* **61**, 925–934.

Nephew, B. C., Kahn, S. A. & Romero, L. M. 2003. Heart rate and behavior are regulated independently of corticosterone following diverse acute stressors. *Gen. Comp. Endocrinol.* **133**, 173–180.

Newton, I. 1998. *Population limitation in birds*. San Diego: Academic Press.

Nonacs, P. 2001. State dependent patch use and the Marginal Value Theorem. *Behav. Ecol.* **12**, 71–83.

Nonacs, P. & Blumstein, D. T. 2010. Predation risk and behavioral life history. In *Evolutionary behavioral ecology*, D. F. Westneat & C. W. Fox (eds.). New York: Oxford University Press.

Nowacek, D. P., Thorne, L. H., Johnston, D. W. & Tyack, P. L. 2007. Responses of cetaceans to anthropogenic noise. *Mammal. Rev.* **37**, 81–115.

NRC. 2007. *Environmental impacts of wind-energy projects*. Washington, DC: National Academies Press.

Nunn, C. L. & Barton, R. A. 2001. Comparative methods for studying primate adaptation and allometry. *Evol. Anthropol.* **10**, 81–98.

Olsson, O., Brown, J. S. & Smith, H. G. 2002. Long- and short-term state-dependent foraging under predation risk: An indication of habitat quality. *Anim. Behav.* **63**, 981–989.

Oro, D. 2008. Living in a ghetto within a local population: An empirical example of an ideal despotic distribution. *Ecology* **89**, 838–846.

Osenberg, C. W., Sarnelle, O. & Cooper, S. D. 1997. Effect size in ecological experiments: The application of biological models in meta-analysis. *Am. Nat.* **150**, 798–812.

Ostro, L. E. T., Silver, S. C., Koontz, F. W., Horwich, R. H. & Brockett, R. 2001. Shifts in social structure of black howler (*Alouatta pigra*) groups associated with natural and experimental variation in population density. *Int. J. Primatol.* **22**, 733–748.

Palagi, E. & Dapporto, L. 2006. Beyond odor discrimination: Demonstrating individual recognition by scent in *Lemur catta*. *Chem. Senses* **31**, 437–443.

Palomino, D. & Carrascal, L. M. 2007. Threshold distances to nearby cities and roads influence the bird community of a mosaic landscape. *Biol. Conserv.* **140**, 100–109.

Parker, P. G. & Waite, T. A. 1997. Mating systems, effective population size, and conservation of natural populations. In *Behavioral approaches to conservation in the wild*, R. Buchholz & J. R. Clemmons (eds.), pp. 243–261. Cambridge: Cambridge University Press.

Parmigiani, S. & vom Saal, F. S. (eds.) 1994. *Infanticide and parental care*. Chur, Switzerland: Harwood Academic Publishers.

Parmigiani, S., Palanza, P. & Vom Saal, F. S. 1998. Ethotoxicology: An evolutionary approach to the study of environmental endocrine-disrupting chemicals. *Toxicol. Ind. Health* **14**, 333–339.

Partan, S. R. & Marler, P. 2005. Issues in the classification of multimodal communication signals. *Am. Nat.* **166**, 231–245.

Partan, S. R., Larco, C. P. & Owens, M. J. 2009. Wild tree squirrels respond with multisensory enhancement to conspecific robot alarm behaviour. *Anim. Behav.* **77**, 1127–1135.

Patricelli, G. L. & Blickley, J. L. 2006. Overview: Avian communication in urban noise: The causes and consequences of vocal adjustment. *Auk* **123**, 639–649.

Patricelli, G. L., Uy, A. C., Walsh, G. & Borgia, G. 2002. Male displays adjusted to female's response. *Nature* **415**, 279–280.

Patricelli, G. L., Coleman, S. W. & Borgia, G. 2006. Male satin bowerbirds (*Ptilonorhynchus violaceus*) adjust their display intensity in response to female startling: An experiment with robotic females. *Anim. Behav.* **71**, 49–59.

Piatt, J. F., Roberts, B. D., Lidster, W. W., Wells, J. L. & Hatch, S. A. 1990. Effects of human disturbance on breeding least and crested auklets at St. Lawrence Island, Alaska. *Auk* **107**, 342–350.

Pollard, K. A. & Blumstein, D. T. 2010. Social group size drives the evolution of individuality. *Unpublished Manuscript*.

Powell, A. N. & Collier, C. L. 1998. Reproductive success of Belding's Savannah sparrows in a highly fragmented landscape. *Auk* **115**, 508–513.

Preisser, E. L., Bolnick, D. I. & Benard, M. F. 2005. Scared to death? The effects of intimidation and consumption in predator–prey interactions. *Ecology* **86**, 501–509.

Price, M. V. & Correll, R. A. 2001. Depletion of seed patches by Merriam's kangaroo rats: Are GUD assumptions met? *Ecol. Lett.* **4**, 334–343.

Primack, R. B. 2008. *A primer of conservation biology*. 4th ed. Sunderland, MA: Sinauer Associates.

Pruett-Jones, S. G. & Lewis, M. J. 1990. Sex ratio and habitat limitation promote delayed dispersal in superb fairy-wrens. *Nature* **348**, 541–542.

Pulliam, H. R. 1973. On the advantages of flocking. *J. Theor. Biol.* **38**, 419–422.

Pullin, A. S. & Knight, R. L. 2009. Doing more good than harm: Building an evidence-base for conservation and environmental management. *Biol. Conserv.* **142**, 931–934.

Purvis, A. & Rambaut, A. 1995. Comparative analysis by independent contrasts (CAIC): An Apple Macintosh application for analysing comparative data. *Comp. Appl. Biosci.* **11**, 247–251.

Pytte, C. L., Rusch, K. M. & Ficken, M. S. 2003. Regulation of vocal amplitude by the blue-throated hummingbird, *Lampornis clemenciae*. *Anim. Behav.* **66**, 703–710.

Quinn, J. L., Whittingham, M. J., Butler, S. J. & Cresswell, W. 2006. Noise, predation risk compensation and vigilance in the chaffinch *Fringilla coelebs*. *J. Avian Biol.* **37**, 601–608.

Rabin, L. A., McCowan, B., Hooper, S. L. & Owings, D. H. 2003. Anthropogenic noise and its effects on animal communication: An interface between comparative psychology and conservation biology. *Int. J. Comp. Psychol.* **16**, 172–192.

Ramsey, D., Spencer, N., Caley, P., Efford, M., Hansen, K., Lam, M. & Cooper, D. 2002. The effects of reducing population density on contact rates between brushtail possums: Implications for transmission of bovine tuberculosis. *J. Appl. Ecol.* **39**, 806–818.

Rannala, B. & Brown, C. R. 1994. Relatedness and conflict over optimal group-size. *Trends Ecol. Evol.* **9**, 117–119.

Réale, D. & Festa-Bianchet, M. 2003. Predator-induced natural selection on temperament in bighorn ewes. *Anim. Behav.* **65**, 463–470.

Réale, D., Gallant, B. Y., Leblanc, M. & Festa-Bianchet, M. 2000. Consistency of temperament in bighorn ewes and correlates with behaviour and life history. *Anim. Behav.* **60**, 589–597.

Réale, D., Reader, S. M., Sol, D., McDougall, P. T. & Dingemanse, N. J. 2007. Integrating animal temperament within ecology and evolution. *Biol. Rev.* **82**, 1–28.

Reed, J. M. 1999. The role of behavior in recent avian extinctions and endangerments. *Conserv. Biol.* **13**, 232–241.

Reid, M. L. & Stamps, J. A. 1997. Female mate choice tactics in a resource-based mating system: Field tests of alternative models. *Am. Nat.* **150**, 98–121.

Reid, W. V. & Miller, K. R. 1989. *Keeping options alive: The scientific basis for conserving biodiversity*. Washington, DC: World Resources Institute.

Reijnen, R., Foppen, R., Braak, C. T. & Thissen, J. 1995. The effects of car traffic on breeding bird populations in woodland. III. Reduction of density in relation to the proximity of main roads. *J. Appl. Ecol.* **32**, 187–202.

Reimchen, T. E. 1994. Predators and morphological evolution in threespine stickleback. In *The evolutionary biology of the threespine stickleback*, M. A. Bell & S. A. Foster (eds.), pp. 240–276. Oxford: Oxford University Press.

Richardson, C. T. & Miller, C. K. 1997. Recommendations for protecting raptors from human disturbance: A review. *Wildl. Soc. Bull.* **25**, 634–638.

Ripple, W. J. & Beschta, R. L. 2004. Wolves and the ecology of fear: Can predation risk structure ecosystems? *BioScience* **54**, 755–766.

Roberts, S. C. & Gosling, L. M. 2004. Manipulation of olfactory signaling and mate choice for conservation breeding: A case study of harvest mice. *Conserv. Biol.* **18**, 548–556.

Robertson, B. A. & Hutto, R. L. 2006. A framework for understanding ecological traps and an evaluation of existing evidence. *Ecology* **87**, 1075–1085.

Robertson, B. C., Elliott, G. P., Eason, D. K., Clout, M. N. & Gemmell, N. J. 2006. Sex allocation theory aids species conservation. *Biol. Lett.* **2**, 229–231.

Rodenhouse, N. L., Sillett, T. S., Doran, P. J. & Holmes, R. T. 2003. Multiple density-dependence mechanisms regulate a migratory bird population during the breeding season. *Proc. R. Soc. B Biol. Sci.* **270**, 2105–2110.

Rodgers, J. A., Jr. & Schwikert, S. T. 2002. Buffer-zone distance to protect foraging and loafing waterbirds from disturbance by personal watercraft and outboard-powered boats. *Conserv. Biol.* **16**, 216–224.

Rodgers, J. A., Jr. & Smith, H. T. 1995. Set-back distances to protect nesting bird colonies from human disturbance in Florida. *Conserv. Biol.* **9**, 89–99.

Rodríguez-Prieto, I. & Fernández-Juricic, E. 2005. Effects of direct human disturbance on the endemic Iberian frog *Rana iberica* at individual and population levels. *Biol. Conserv.* **123**, 1–9.

Roemer, G. W., Donlan, C. J. & Courchamp, F. 2002. Golden eagles, feral pigs and insular carnivores: How exotic species turn native predators into prey. *Proc. Natl. Acad. Sci. USA* **99**, 791–796.

Rohlf, F. J. & Sokal, R. R. 1994. *Statistical tables*. 3rd ed. San Francisco: W. H. Freeman.

Ronconi, R. A. & St. Clair, C. C. 2006. Efficacy of a radar-activated on-demand system for deterring waterfowl from oil sands tailing ponds. *J. Appl. Ecol.* **43**, 111–119.

Ropeik, D. & Gray, G. 2002. *Risk: A practical guide for deciding what's really safe and what's really dangerous in the world around you*. Boston: Houghton Mifflin.

Rosenthal, R. 1991. *Meta-analytic procedures for social research*. Newbury Park, CA: Sage Publications.

Roughgarden, J. 1972. Evolution of niche width. *Am. Nat.* **106**, 683–718.

Rowe, N. 1996. *The pictorial guide to the living primates*. Charlestown, RI: Pogonias Press.

Ryan, M. J. & Keddy-Hector, A. 1992. Directional patterns of female mate choice and the role of sensory biases. *Am. Nat.* **139**, S4–S35.

Sandin, S. A. & Pacala, S. W. 2005. Fish aggregation results in inversely density-dependent predation on continuous coral reefs. *Ecology* **86**, 1520–1530.

Sayer, J. A. 1991. *Rainforest buffer zones: Guidelines for protected area managers*. Gland, Switzerland: IUCN.

Scheifele, P. M., Andrew, S., Cooper, R. A., Darre, M., Musiek, F. E. & Max, L. 2005. Indication of a Lombard vocal response in the St. Lawrence River beluga. *J. Acoust. Soc. Am.* **117**, 1486–1492.

Schlaepfer, M. A., Runge, M. C. & Sherman, P. W. 2002. Ecological and evolutionary traps. *Trends Ecol. Evol.* **17**, 474–480.

Schlossberg, S. R. & Ward, M. P. 2004. Using conspecific attraction to conserve endangered birds. *Endang. Sp. Update* **21**, 132–138.

Schlupp, I. & Podloucky, R. 1994. Changes in breeding site fidelity: A combined study of conservation and behaviour in the common toad *Bufo bufo*. *Biol. Conserv.* **69**, 285–291.

Schlupp, I., Marler, C., & Ryan, M. J. 1994. Benefit to male sailfin mollies of mating with heterospecific females. *Science* **263**, 373–374.

Schorger, A. W. 1955. *The passenger pigeon: Its natural history and extinction*. Norman: University of Oklahoma Press.

Schulte, B. A., Bagley, K. R., Groover, M., Loizi, H., Merte, C., Meyer, J. M., Napora, E., Stanley, L., Vyas, D. K., Wollett, K., Goodwin, T. E. & Rasmussen, L. E. L. 2007. Comparisons of state and likelihood of performing chemosensory event behaviors in two populations of African elephants (*Loxodonta africana*). In *Chemical signals in vertebrates 11*, R. Beynon, J. Hurst, C. Roberts & T. Wyatt (eds.), pp. 81–90. New York: Springer.

Scott, D. K. 1988. Breeding success in Bewick's swans. In *Reproductive success: Studies of individual variation in contrasting breeding systems*, T. H. Clutton-Brock (ed.), pp. 220–236. Chicago: University of Chicago Press.

Seddon, P. J., Armstrong, D. P. & Maloney, R. F. 2007. Developing the science of reintroduction biology. *Conserv. Biol.* **21**, 303–312.

Seehausen, O., van Alphen, J. M. & Witte, F. 1997. Cichlid fish diversity threatened by eutrophication that curbs sexual selection. *Science* **277**, 1808–1811.

Seppänen, J.-T., Forsman, J. T., Mönkkönen, M. & Thomson, R. L. 2007. Social information use

is a process across space, time and ecology, reaching heterospecifics. *Ecology* **88**, 1622–1633.

Shafer, C. L. 1999a. National park and reserve planning to protect biological diversity: Some basic elements. *Landsc. Urban Plann.* **44**, 123–153.

Shafer, C. L. 1999b. U.S. national park buffer zones: Historical, scientific, social and legal aspects. *Environ. Manag.* **23**, 49–73.

Shelley, E. L. & Blumstein, D. T. 2005. The evolution of vocal alarm communication in rodents. *Behav. Ecol.* **16**, 169–177.

Shettleworth, S. J. 1998. *Cognition, evolution and behavior*. New York: Oxford University Press.

Shier, D. M. 2005. Translocations are more successful when prairie dogs are moved as families. In *Conservation of the black-tailed prairie dog*, J. L. Hoogland (ed.), pp. 189–190. Washington, DC: Island Press.

Shier, D. M. 2006. Effect of family support on the success of translocated black-tailed prairie dogs. *Conserv. Biol.* **20**, 1780–1790.

Shier, D. M. & Owings, D. H. 2006. Effects of predator training on behavior and post-release survival of captive prairie dogs (*Cynomys ludovicianus*). *Biol. Conserv.* **132**, 126–135.

Shier, D. M. & Owings, D. H. 2007a. Effects of social learning on predator training and postrelease survival in juvenile black-tailed prairie dogs, *Cynomys ludovicianus*. *Anim. Behav.* **73**, 567–577.

Shier, D. & Owings, D. 2007b. Social influences on predator training for conservation. *Conserv. Behav.* **5**, 6–8.

Shochat, E., Patten, M. A., Morris, D. W., Reinking, D. L., Wolfe, D. H. & Sherrod, S. K. 2005. Ecological traps in isodars: Effects of tall-grass prairie management on bird nest success. *Oikos* **111**, 159–169.

Short, J., Kinnear, J. E. & Robley, A. 2002. Surplus killing by introduced predators in Australia: Evidence for ineffective anti-predator adaptations in native prey species? *Biol. Conserv.* **103**, 283–301.

Shustack, D. P. & Rodewald, A. D. 2008. Understanding demographic and behavioral mechanisms that guide responses of Neotropical migratory birds to urbanization: A simulation approach. *Avian Conserv. Ecol.* **2(3)**, 2. www.ace-eco.org/vol3/iss2/art2.

Sibly, R. M. 1983. Optimal group size is unstable. *Anim. Behav.* **31**, 947–948.

Sih, A. 1984. Optimal behavior and density-dependent predation. *Am. Nat.* **123**, 314–326.

Sih, A., Bell, A. M. & Johnson, J. C. 2004a. Behavioral syndromes: An ecological and evolutionary overview. *Trends Ecol. Evol.* **19**, 372–378.

Sih, A., Bell, A. M., Johnson, J. C. & Ziemba, R. E. 2004b. Behavioral syndromes: An integrative overview. *Q. Rev. Biol.* **79**, 241–277.

Sinn, D. L., Apiolaza, L. A. & Moltschaniwsky, N. A. 2006. Heritability and fitness-related consequences of squid personality traits. *J. Evol. Biol.* **19**, 1437–1447.

Skagen, S. K., Knight, R. L. & Orians, G. H. 1991. Human disturbance of an avian scavenging guild. *Ecol. Appl.* **1**, 215–225.

Slabbekoorn, H. & den Boer-Visser, A. 2006. Cities change the songs of birds. *Curr. Biol.* **16**, 2326–2331.

Slabbekoorn, H. & Peet, M. 2003. Birds sing at a higher pitch in urban noise. *Nature* **424**, 267.

Slos, S. & Stoks, R. 2006. Behavioural correlations may cause partial support for the risk allocation hypothesis in damselfly larvae. *Ethology* **112**, 143–151.

Smith, B. R. & Blumstein, D. T. 2008. Fitness consequences of personality: A meta-analysis. *Behav. Ecol.* **19**, 448–455.

Smith, C., Reynolds, J. D. & Sutherland, W. J. 2000. Population consequences of reproductive decisions. *Proc. R. Soc. Lond. B Biol. Sci.* **267**, 1327–2334.

Smith, D. W., Peterson, R. O. & Houston, D. B. 2004. Yellowstone after wolves. *BioScience* **53**, 330–340.

Smith, G. C. 2001. Models of *Mycobacterium bovis* in wildlife and cattle. *Tuberulosis* **81**, 51–64.

Snyder, N. F. R., Derrickson, S. R., Beissinger, S. R., Wiley, J. W., Smith, T. B., Toone, W. D. & Miller, B. 1996. Limitations of captive breeding in endangered species recovery. *Conserv. Biol.* **10**, 338–348.

Sodhi, N. S. 2002. Competition in the air: Bird versus aircraft. *Auk* **119**, 587–595.

Sokal, R. R. & Rohlf, F. J. 1981. *Biometry: The principles and practice of statistics in biological research.* 3rd ed. New York: W. H. Freeman.

Solomon, N. G. & French, J. A. 1997. *Cooperative breeding in mammals.* Cambridge: Cambridge University Press.

Soulé, M. E. & Wilcox, B. (eds.) 1980. *Conservation biology: An evolutionary-ecological perspective.* Sunderland, MA: Sinauer Associates.

Soulé, M. E., Alberts, A. C. & Bolger, D. T. 1992. The effects of habitat fragmentation on chaparral plants and vertebrates. *Oikos* **63**, 39–47.

St. Clair, C. C., Bélisle, M., Desrochers, A. & Hannon, S. 1998. Winter responses of forest birds to habitat corridors and gaps. *Conserv. Ecol.* **2(2)**, 13, www.consecol.org/vol2/iss2/art13.

Stamps, J. A. 1988. Conspecific attraction and aggregation in territorial species. *Am. Nat.* **131**, 329–347.

Stamps, J. A. 2001. Habitat selection by dispersers: Integrating proximate and ultimate approaches. In *Dispersal*, J. Clobert, E. Danchin, A. A. Dhondt & J. D. Nichols (eds.), pp. 230–242. New York: Oxford University Press.

Stankowich, T. & Blumstein, D. T. 2005. Fear in animals: A meta-analysis and review of risk assessment. *Proc. R. Soc. B Biol. Sci.* **272**, 2627–2634.

Stankowich, T. & Coss, R. G. 2007. The re-emergence of felid camouflage with the decay of predator recognition in deer under relaxed selection. *Proc. R. Soc. B Biol. Sci.* **274**, 175–182.

Steele, B. M. & Hogg, J. T. 2003. Measuring individual quality in conservation and behavior studies. In *Animal behavior and wildlife conservation*, M. Festa-Bianchet & M. Apollonio (eds.). Washington, DC: Island Press.

Stephens, D. W., Brown, J. S. & Ydenberg, R. C. (eds.) 2007. *Foraging: Behavior and ecology.* Chicago: University of Chicago Press.

Strier, K. B. 1997. Behavioral ecology and conservation biology of primates and other animals. *Adv. Study Behav.* **26**, 101–158.

Sugg, D. W. & Chesser, R. K. 1994. Effective population sizes with multiple paternity. *Genetics* **137**, 1147–1155.

Sugg, D. W., Chesser, R. K., Dobson, F. S. & Hoogland, J. L. 1996. Population genetics meets behavioral ecology. *Trends Ecol. Evol.* **11**, 338–343.

Sutherland, W. J. 1996. *From individual behaviour to population ecology.* Oxford: Oxford University Press.

Sutherland, W. J. 1998a. The effect of local change in habitat quality on populations of migratory species. *J. Appl. Ecol.* **35**, 418–421.

Sutherland, W. J. 1998b. The importance of behavioural studies in conservation biology. *Anim. Behav.* **56**, 801–809.

Sutherland, W. J., Pullin, A. S., Dolman, P. M. & Knight, T. M. 2004. The need for evidence-based conservation. *Trends Ecol. Evol.* **19**, 305–308.

Svendsen, G. E. & Armitage, K. B. 1973. Mirror-image stimulation applied to field behavioral studies. *Ecology* **54**, 623–627.

Swaisgood, R. R. 2007. Current status and future directions of applied behavioral research for animal welfare and conservation. *Appl. Anim. Behav. Sci.* **102**, 139–162.

Swaisgood, R. R., White, A. M., Zhou, X., Zhang, H., Zhang, G., Wei, R., Hare, V. J., Tepper, E. M. & Lindburg, D. G. 2001. A quantitative assessment of the efficacy of an environmental enrichment programme for giant pandas. *Anim. Behav.* **61**, 447–457.

Swaisgood, R. R., Dickman, D. M. & White, A. M. 2006. A captive population in crisis: Testing hypotheses for reproductive failure in captive-born southern white rhinoceros females. *Biol. Conserv.* **129**, 468–476.

Swenson, J. E., Sandegren, F., Söderberg, A., Bjärvall, A., Franzén, R. & Wabakken, P. 1997. Infanticide caused by hunting of male bears. *Nature* **386**, 450–451.

Tang-Martinez, Z. & Ryder, T. B. 2005. The problem with paradigms: Bateman's worldview as a case study. *Integr. Comp. Biol.* **45**, 821–830.

Tarlow, E. M. & Blumstein, D. T. 2007. Evaluating methods to quantify anthropogenic stressors on wild animals. *Appl. Anim. Behav. Sci.* **102**, 429–451.

Taylor, P. D., Fahrig, L., Henein, K. & Merriam, G. 1993. Connectivity is a vital element of landscape structure. *Oikos* **68**, 571–573.

Templeton, J. J. & Giraldeau, L.-A. 1995. Patch assessment in foraging flocks of European starlings: Evidence for the use of public information. *Behav. Ecol.* **6**, 65–72.

Thornhill, R. 1984. Alternative female choice tactics in the scorpionfly *Hylobittacus apicalis* (Mecoptera) and their implications. *Am. Zool.* **24**, 367–383.

Tinbergen, N. 1963. On aims and methods of ethology. *Z. Tierpsychol.* **20**, 410–433.

Tinker, M. T., Bentall, G. & Estes, J. A. 2008. Food limitation leads to behavioral diversification and dietary specialization in sea otters. *Proc. Natl. Acad. Sci. USA* **105**, 560–565.

Tisdale, V. & Fernández-Juricic, E. 2009. Vigilance and predator detection vary between avian species with different visual acuity and coverage. *Behav. Ecol.* **20**, 936–945.

Treves, A. 2000. Theory and method in studies of vigilance and aggregation. *Anim. Behav.* **60**, 711–722.

Trivers, R. L. & Willard, D. E. 1973. Natural selection of parental ability to vary the sex ratio of offspring. *Science* **179**, 90–92.

Turchin, P. 1995. Population regulation: Old arguments and a new synthesis. In *Population dynamics*, N. Cappuccino & P. Price (eds.), pp. 19–40. New York: Academic Press.

USFWS. 2006. *Draft recovery plan for the black-footed ferret (Mustela nigripes)*. Denver: Region 6, US Fish and Wildlife Service.

USFWS. 2007. *Red wolf* (Canis rufus) *5-year status review: Summary and evaluation*. Manteo, NC: US Fish and Wildlife Service.

Valone, T. J. 2007. From eavesdropping on performance to copying the behavior of others: a review of public information use. *Behav. Ecol. Sociobiol.* **62**, 1–14.

van der Merwe, M. & Brown, J. S. 2008. Mapping the landscape of fear of the Cape ground squirrel (*Xerus inauris*). *J. Mammal.* **89**, 1162–1169.

Van Horne, B. 1983. Density as a misleading indicator of habitat quality. *J. Wildl. Manag.* **47**, 893–901.

van Oers, K., de Jong, G., Drent, P. J. & van Noordwijk, A. J. 2004. A genetic analysis of avian personality traits: Correlated, response to artificial selection. *Behav. Genet.* **34**, 611–619.

van Schaik, C. P. & Janson, C. H. (eds.) 2000. *Infanticide by males and its implications*. Cambridge: Cambridge University Press.

Village, A. 1990. *The kestrel*. London: T. & A. D. Poyser.

Vine, I. 1971. Risk of visual detection and pursuit by a predator and the selective advantage of flocking behaviour. *J. Theor. Biol.* **30**, 405–422.

von Uexküll, J. 1934. A stroll through the worlds of animals and men. In *Instinctive behaviour*, C. H. Schiller (ed.), pp. 5–15. London: Methuen.

Vos, D. K., Ryder, R. A. & Graul, W. D. 1985. Response of breeding great blue herons to human disturbance in northcentral Colorado. *Colo. Waterbirds* **8**, 13–22.

Wagner, W. E., Jr. 1998. Measuring female mating preferences. *Anim. Behav.* **55**, 1029–1042.

Walters, C. J. & Holling, C. S. 1990. Large-scale management experiments and learning by doing. *Ecology* **71**, 2060–2068.

Walters, J. R., Copeyton, C. K. & Carter, J. H., III. 1992. Test of the ecological basis of cooperative breeding in red-cockaded woodpeckers. *Auk* **109**, 90–97.

Walther, G.-R., Roques, A., Hulme, P. E., Sykes, M. T., Pysek, P., Kühn, I., Zobel, M., Bacher, S., Botta-Dukát, Z., Bugmann, H., Czúcz, B., Dauber, J., Hickler, T., Jarosík, V., Kenis, M., Klotz, S., Minchin, D., Moora, M., Nentwig, W., Ott, J., Panov, V. E., Reineking, B., Robinet, C., Semenchenko, V., Solarz, W., Thuiller, W., Vilà, M., Vohland, K. & Settele, S. 2009. Alien species in a warmer world: Risks and opportunities. *Trends Ecol. Evol.* **24**, 686–693.

Ward, M. P. & Schlossberg, S. R. 2004. Conspecific attraction and the conservation of territorial songbirds. *Conserv. Biol.* **18**, 519–525.

Warren, P. S., Katti, M., Ermann, M. & Brazel, A. 2006. Urban bioacoustics: It's not just noise. *Anim. Behav.* **71**, 491–502.

Watters, J. V. & Meehan, C. L. 2007. Different strokes: Can managing behavioral types increase post-release success. *Appl. Anim. Behav. Sci.* **102**, 364–379.

Werner, S. J., Carlson, J. C., Tupper, S. K., Santer, M. M. & Linz, G. M. 2009. Threshold concentrations of an anthraquinone-based repellent for Canada geese, red-winged blackbirds, and ring-necked pheasants. *Appl. Anim. Behav.* **121**, 190–196.

West, A. D., Goss-Custard, J. D. , Stillman, R. A., Caldow, R. W. G., le V. dit Durell, S. E. A. & McGrorty, S. 2002. Predicting the impacts of disturbance on shorebird mortality using a behaviour-based model. *Biol. Conserv.* **106**, 319–328.

Wey, T., Blumstein, D. T., Shen, W. & Jordán, F. 2008. Social network analysis of animal behaviour: A promising tool for the study of sociality. *Anim. Behav.* **75**, 333–344.

White, A. M., Swaisgood, R. R. & Zhang, H. 2002. The highs and lows of chemical communication in giant pandas (*Ailuropoda melanoleuca*): Effect of scent deposition height on signal discrimination. *Behav. Ecol. Sociobiol.* **51**, 519–529.

White, G. C. & Burnham, K. P. 1999. Program MARK: Survival estimation from populations of marked animals. *Bird Study* **46S**, 120–138.

Whitehead, H. 2008. *Analyzing animal societies: Quantitative methods for vertebrate social analysis.* Chicago: University of Chicago Press.

Whitman, K., Starfield, A. M., Quadling, H. S. & Packer, C. 2004. Sustainable trophy hunting of African lions. *Nature* **428**, 175–178.

Wiley, E. O., Siegel-Causey, D., Brooks, D. R. & Funk, V. A. 1991. *The compleat cladist.* Special Publication. Lawrence: University of Kansas.

Williams, B. K., Nichols, J. D. & Conroy, M. J. 2002. *Analysis and management of animal populations.* San Diego: Academic Press.

Williams, C. K., Lutz, R. S. & Applegate, R. D. 2003. Optimal group size and northern bobwhite coveys. *Anim. Behav.* **66**, 377–387.

Wilson, D. S. 1998. Adaptive individual differences within single populations. *Philos. Trans. R. Soc. Lond. B* **353**, 199–205.

Winer, B. J. 1962. *Statistical principles in experimental design.* New York: McGraw-Hill.

Wingfield, J. C. & Sapolsky, R. M. 2003. Reproduction and resistance to stress: When and how. *J. Neuroend.* **15**, 711–724.

Wirtz, P. & Lörscher, J. 1983. Group sizes of antelopes in an East African National Park. *Behaviour* **84**, 135–156.

Wittenberger, J. F. 1983. Tactics of mate choice. In *Mate choice*, P. Bateson (ed.), pp. 435–447. Cambridge: Cambridge University Press.

Wolf, C. M., Griffith, B., Reed, C. & Temple, S. A. 1996. Avian and mammalian translocations: Update and reanalysis of 1987 survey data. *Conserv. Biol.* **10**, 1142–1154.

Wolf, C. M., Garland, T. J. & Griffith, B. 1998. Predictors of avian and mammalian translocation success: Reanalysis with phylogenetically independent contrasts. *Biol. Conserv.* **86**, 243–255.

Wood, W. E. & Yezerinac, S. M. 2006. Song sparrow (*Melospiza melodia*) song varies with urban noise. *Auk* **123**, 650–659.

Woodroffe, R., Donnelly, C. A., Cox, D. R., Bourne, F. J., Cheeseman, C. L., Delahay, R. J., Gettinby, G., McInerney, J. P. & Morrison, W. I. 2005. Effects of culling on badger (*Meles meles*) spatial organization: Implications for the control of bovine tuberculosis. *J. Appl. Ecol.* **43**, 1–10.

Wright, S. 1938. Size of population and breeding structure in relation to evolution. *Science* **87**, 430–431.

Wright, S. 1969. *Evolution and the genetics of populations.* Vol. 2. *The theory of gene frequencies.* Chicago: University of Chicago Press.

Wright, S. 1978. *Evolution and the genetics of populations.* Vol. 4. *Variability within and among natural populations.* Chicago: University of Chicago Press.

Wroe, S., Field, J., Fullagar, R. & Jermiin, L. S. 2004. Megafaunal extinction in the Late Quaternary and the global overkill hypothesis. *Alcheringa* **28**, 291–331.

Zemel, A. & Lubin, Y. 1995. Inter-group competition and stable group sizes. *Anim. Behav.* **50**, 485–488.

Zollner, P. A. & Lima, S. L. 1997. Landscape-level perceptual abilities in white-footed mice: Perceptual range and the detection of forested habitat. *Oikos* **80**, 51–60.

Zollner, P .A. & Lima, S. L. 2005. Behavioral trade-offs when dispersing across a patchy landscape. *Oikos* **108**, 219–230.

Zuk, M., Johnson, K., Thornhill, R. & Ligon, J. D. 1990. Mechanisms of female choice in red jungle fowl. *Evolution* **44**, 477–485.

Index